Surface-Enhanced Vibrational Spectroscopy

Surface-Enhanced Vibrational Spectroscopy

Ricardo Aroca
University of Windsor, Ontario, Canada

John Wiley & Sons, Ltd

Telephone (+44) 1243 779777

Email (for orders and customer service enquiries): cs-books@wiley.co.uk
Visit our Home Page on www.wileyeurope.com or www.wiley.com

Other Wiley Editorial Offices

John Wiley & Sons Inc., 111 River Street, Hoboken, NJ 07030, USA

Jossey-Bass, 989 Market Street, San Francisco, CA 94103-1741, USA

Wiley-VCH Verlag GmbH, Boschstr. 12, D-69469 Weinheim, Germany

John Wiley & Sons Australia Ltd, 42 McDougall Street, Milton, Queensland 4064, Australia

John Wiley & Sons (Asia) Pte Ltd, 2 Clementi Loop #02-01, Jin Xing Distripark, Singapore 129809

John Wiley & Sons Canada Ltd, 22 Worcester Road, Etobicoke, Ontario, Canada M9W 1L1

Wiley also publishes its books in a variety of electronic formats. Some content that appears in print may not be available in electronic books.

Library of Congress Cataloging-in-Publication Data

Aroca, Ricardo.
Surface enhanced vibrational spectroscopy / Ricardo Aroca.
p. cm.
Includes bibliographical references and index.
ISBN-13: 978-0-471-60731-1 (acid-free paper)
ISBN-10: 0-471-60731-2 (acid-free paper)
1. Vibrational spectra. 2. Molecular spectroscopy. 3. Raman effect, Surface enhanced.
I. Title.
QD96.V53A76 2006
543′.54–dc22

2005036662

British Library Cataloguing in Publication Data

A catalogue record for this book is available from the British Library

ISBN-13 978-0-471-60731-1
ISBN-10 0-471-60731-2

Typeset in 10.5/13pt Sabon by TechBooks, New Delhi, India
Printed and bound in Great Britain by TJ International, Padstow, Cornwall
This book is printed on acid-free paper responsibly manufactured from sustainable forestry in which at least two trees are planted for each one used for paper production.

ac

To my wife Patricia, our children: Patricia Paulina, Marcela Susana and Ricardo Andres, and our grandchildren: Miguel, Stéphane, Natalia, Madison, Callum and Maria Elena

Contents

Preface

Everything is vague to a degree you do not realize till you have tried to make it precise.

Bertrand Russell
British author, mathematician, and philosopher (1872–1970)

Surface-enhanced Raman scattering (SERS) is a moving target. Every time you look at it, it mutates, and new speculations are suddenly on the horizon. This elusiveness seems to defy our ability to predict the outcome of each new SERS experiment. The uncertainty is even more challenging when one approaches the single molecule regime (single molecule detection – SMD), since the attempt at experimental measurement of SERS may actually affect the molecule, or the nanostructures interacting with the molecular system, or both. However, one should not be surprised by this lack of determinism. While it is often taken for granted in the analytical spectroscopy of ensemble averages, it is particularly significant in ultrasensitive chemical analysis where one is dealing with only a few quantum systems (molecules) and nanostructures, with pronounced quantum effects. The difficulty is compounded by the fact that the enhanced signal is the result of several contributions, and their separation into well-defined components is virtually impossible. Observed SERS spectra are the final result of multiple factors, and the contribution of these factors is case specific. It is therefore of the utmost importance to examine and analyze closely the set of variables that may play a role in producing observed SERS spectra.

In this book, SERS is narrowly defined as surface *plasmon-assisted enhancement* of Raman scattering. Therefore, the term SERS is used for molecules located on, or close to, nanostructures that can support surface plasmons leading to an electromagnetic (EM) field enhancement of the Raman signal. This definition excludes smooth surfaces with only nonradiative plasmons and small atomic clusters where surface plasmons are not realized. There is consensus on the electromagnetic origin and fundamental properties of the signal enhancement of SERS, as assisted by surface plasmon excitation on certain nanostructures. Thus, the presence of this component in the observed enhanced intensity will define the observed spectrum as a SERS spectrum.

Defining SERS in terms of one of the components of the observed enhanced intensity may, at first, seem limited and narrow. However, this definition provides the basis for a full discussion of the observations, and also a guide for the experimentalist to tune experimental conditions according to the ultimate goal of their research project. The definition does not necessarily imply that the plasmon assisted contribution ought to be the largest; other resonances may contribute and, in some cases, produce dominant contributions. However, it is the presence of the plasmon resonance that will define the observed spectral intensities as a SERS spectrum. In addition to this binding definition, the main thrust of the book is to discuss only two of the many enhanced optical phenomena in surface-enhanced spectroscopy: surface-enhanced Raman scattering (SERS) and surface-enhanced infrared absorption (SEIRA). SERS and SEIRA form a new branch of vibrational spectroscopy, which we now call *surface-enhanced vibrational spectroscopy (SEVS)*, and it serves as the title for the book. SEVS deals with the enhanced spectra of molecules on specially fabricated nanostructures with the ability to support surface plasmons and to enhance optical signals. Stable molecular electronic states are characterized by their vibrational structure [1–3], and the great advantage of vibrational spectroscopy, which can provide the fingerprint of any molecular system, is in the vast body of vibrational assignment data for gas, liquid, solid and, most relevant to SEVS, surface complex systems. SEVS is an extremely powerful addition to surface-sensitive and single molecule spectroscopies (SMS). From the analytical perspective, a concentrated sample of an analyte (the adsorbed molecule to be assessed) should form complete monolayer coverage on the surface plasmon supporting nanostructure. However, SERS and SEIRA are not limited to the first monolayer and, indeed, the EM enhancement is a long-range phenomenon that decays more slowly than the field dipole. That being said, the first layer will dominate the SEVS spectrum, and it is the spectrum of this layer that could be used for the compilation of a database. Ultrasensitive analysis in SERS will start at monolayer coverage and move in the direction of submonolayer coverage, to achieve the ultimate single molecule–nanostructure limit. Selection rules derived for infrared and

Raman spectra [2] also apply to adsorbed species, with some additional qualifications. For highly reflecting surfaces in the infrared region, only those vibrational modes with a component of the dynamic dipole perpendicular to the surface are observed. These stringent 'surface-selection rules' could severely limit the relative intensities in the recorded infrared spectrum. At the same time, this new spectrum provides information on the molecular orientation and molecule–surface interaction. The surface-selection rules that apply to infrared and Raman spectroscopy are extended to SEVS with yet additional qualifications imposed by the nature of the local field and/or the roughness of the surface used for SEVS.

The definition and the main components are illustrated in the cartoon shown in Figure 1, where single particles and clusters of particles supporting surface plasmons are interacting with a molecular probe.

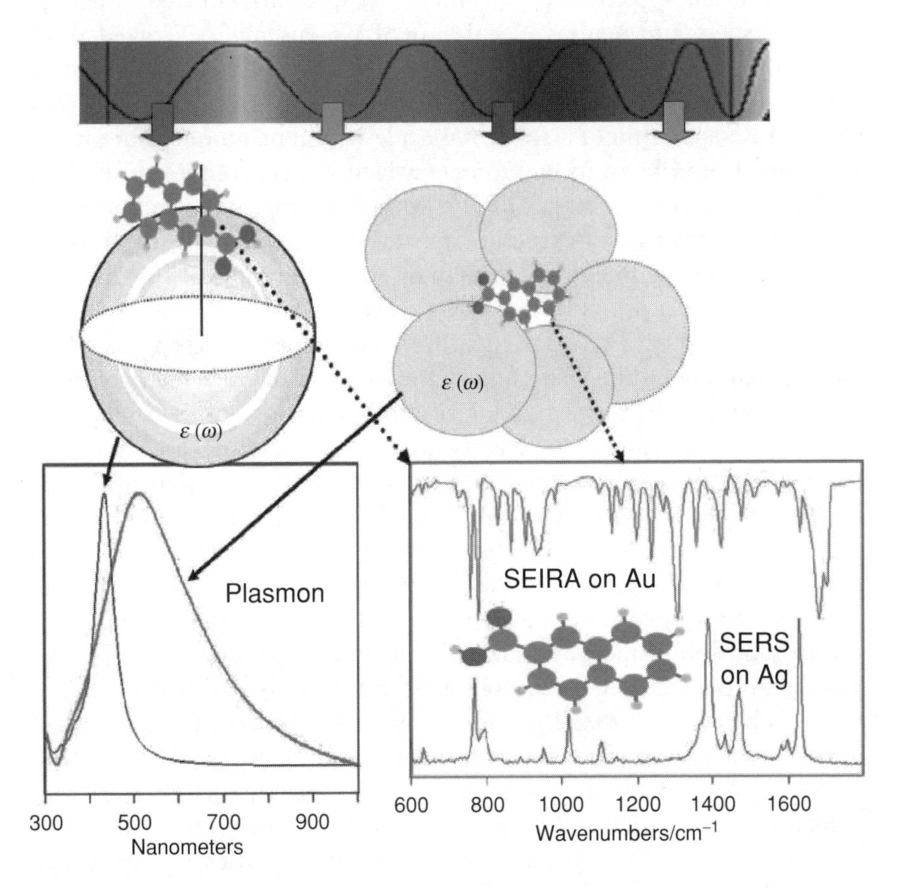

Figure 1 The three SEVS elements: the molecule, the electromagnetic radiation and the nanostructure, with the resulting plasmon and surface-enhanced spectra.

The study of vibrational energy levels, or vibrational spectroscopy, is carried out mainly with infrared absorption or inelastic scattering (Raman) [1–3] of electromagnetic radiation [4]. The quantum description of the vibrating molecule provides the energy levels, and that is followed by the study of the dynamics of the molecule–light interaction [5]. The information obtained from Raman scattering and that gathered from infrared absorption are complementary, to the point of being mutually exclusive for centrosymmetric molecules. SEVS spectra are the result of the molecule–light interaction when the molecule is near or attached to a nanostructure supporting surface plasmons. In the end, regardless of the mechanisms involved, the information, as in vibrational spectroscopy, is contained in a Raman or an infrared spectrum, and the challenge is in the interpretation of these spectra.

In vibrational spectroscopy, the molecular spectra are indeed ensemble-averaged spectra of many molecules. In SEVS, ensemble-averaged SERS and SEIRA spectra also form the bulk of the accumulated spectroscopic data. However, in the absence of the statistical average, the properties of the SERS spectrum of a single molecule would be unique, since it is a very sensitive probe of its environment. Hence it is profitable to make a distinction between 'average SERS' spectra and 'single molecule' spectra.

The presence of a nanostructure, most commonly a noble metal nanostructure, with the intrinsic property of enhancing optical signals, may leave its own footprints in the SEVS spectra. The nanostructure's trail can be detected in a characteristic frequency due to the surface complex, a frequency shift, a peculiar bandwidth, a distinct relative intensity or a temporal behavior giving rise to fluctuations of the signal. Enhancing nanostructures can be fabricated as isolated particles, nanorods, nanowires or aggregates. However, in many applications of SEVS the nanostructures are fabricated on to a solid substrate, and thereby further spectral features may be observed due to reflections and refraction phenomena on the surface of the substrates. Inevitably, there exists the danger of drawing the line in the wrong place when discussing vibrational spectroscopy on surfaces and surface-enhanced vibrational spectroscopy. The definition of SEVS, used here, separates the results obtained on 'flat' or smooth reflecting metal surfaces from the SEVS results obtained on modified surfaces that contain enhancing nanostructures.

In summary, SEVS is the vibrational spectroscopy of molecules that is realized on well-defined nanostructures. It is a new molecular spectroscopy that is highly dependent on the optical properties, size and shape of metallic nanostructures. SERS, in particular, permits giant amplification of the optical signal and single molecule detection. At the SMD level, temporal phenomena or fluctuations may be used as a probe for surface dynamics. Observing and manipulating biomolecules in single molecule spectroscopy may directly reveal their dynamic behavior, knowing that to detect dynamic behavior of target molecules using ensemble-averaged measurements is almost impossible. Experimentally, near-field scanning optical microscopy (NSOM) has joined the common far-field Raman scattering, making it possible to analyze optical properties with a spatial resolution below the diffraction limit. In a parallel development, SEVS is becoming a viable technique for nanoparticle characterization.

This book begins by devoting a chapter to reviewing the vibrating molecule and the origin of infrared and Raman spectra. These are the fundamentals and they provide the reference needed for the interpretation of SEVS results. Chapter 2 contains brief discussions on the absorption and scattering of light by metallic nanoparticles (important for SERS interpretation), the fabrication of nanostructures [6] and the selection of the appropriate experimental conditions for SERS and SEIRA. Light absorption enhancement by nanoparticles and light scattering enhancement by nanoparticles supporting surface plasmon are, in themselves, an active field of research in physics and chemistry [7, 8]. The theory and detection of surface plasmons of isolated particles of different size and shape [9] have been advanced by several groups and the references can be found in Chapter 2. Furthermore, aggregates of nanoparticles can sustain localized and delocalized surface plasmons, and highly localized modes, or *hot spots*, allowing for the concentration of electromagnetic energy in small parts of the system [10]. Finally, a section on reflection spectroscopy with special attention to reflection–absorption infrared spectroscopy (RAIRS) is also included to explain the effect that reflecting surfaces have on the observed relative intensities of vibrational spectra.

Chapter 3 is dedicated to SERS as a surface plasmon-assisted spectroscopy. The most rudimentary models that provide guidance for the experimentalist are also included. Chapter 4 is an attempt to examine the chemical effects, or the role in the observed SERS spectra of contributions due to molecule–nanostructure interactions. Chapter 5 is dedicated to demonstrating that SERS is observed for any type of molecular system, and is, thereby, not molecular specific. A database is provided on the web for the thousands of references that were reviewed. These form a catalog of molecules studied by SERS or SERRS, organized according to the type of molecule system, and intended to help experimentalists who would like to use SERS as an analytical tool. This is not a comprehensive database, but the time has come for the creation of a collection of SERS spectra that will be useful for analytical applications. Chapter 6 is an overview of SERS applications. Chapter 7 describes SEIRA and its applications. Each chapter contains extensive citations to help the user and to make the book a useful reference. The book contains a glossary that is intended to be helpful given the multidisciplinary nature of SERS (chemistry, solid-state physics, optics and electrodynamics).

Thousands of publications, many excellent reviews and, in particular, the expanding analytical applications of SERS and SEIRA are of such importance that there is a need for a text on methods and interpretation of spectra. This book has been written with the intention of meeting, in part, that need. Since much of the material covered in this book is recent, it is not possible to feel as comfortable in the description and of the subject as in a more settled field of spectroscopy, and some users may find the effort premature. However, I believe that the subject dealt with here is important and should be part of the working knowledge of chemists, physicists and material scientists. An attempt to summarize the developments to date is worth the risk of criticism.

REFERENCES

[1] G. Herzberg, *Spectra and Molecular Structure. II. Infrared and Raman Spectra of Polyatomic Molecules*, Van Nostrand, Princeton, NJ, 1945.
[2] E.B. Wilson Jr, J.C. Decius and P.C. Cross, *Molecular Vibrations; The Theory of Infrared and Raman Vibrational Spectra*, McGraw-Hill, New York, 1955.
[3] M.B. Bolkenshtein, L.A. Gribov, M.A. Eliashevich and B.I. Stepanov, *Molecular Vibrations*, Nauka, Moscow, 1972.
[4] M. Born and E. Wolf, *Principles of Optics*, Pergamon Press, Oxford, 1975.
[5] J.D. Macomber, *The Dynamics of Spectroscopic Transitions*, John Wiley & Sons, Inc., New York, 1976.
[6] G.A. Ozin and A.C. Arsenault, *Nanochemistry. A Chemical Approach to Nanomaterials*, Royal Society of Chemistry, Cambridge, 2005.
[7] D.L. Feldheim and C.A. Foss (eds), *Metal Nanoparticles. Synthesis, Characterization and Applications*, Marcel Dekker, New York, 2002.
[8] G. Schmid (ed.), *Nanoparticles. From Theory to Applications*, Wiley-VCH, Weinheim, 2005.
[9] E.A. Coronado and G.C. Schatz, *J. Chem. Phys.*, **119**, 2003, 3926–3934.
[10] M.I. Stockman, S.V. Faleev and D.J. Bergman, *Phys. Rev. Lett.*, 2001, **87**, 167401/1–167401/4.

Acknowledgments

This book is the synergistic product of many people, to whom I extend thanks for their tireless efforts and unique contributions. First, to my students who have worked with me initially at the University of Toronto and then in the Materials and Surface Science Group at the University of Windsor, for their dedication and research that led to many of the ideas in this book. In particular, Paul Goulet, Nicholas Pieczonka and Daniel Ross, who were working with me during the time of writing this book, and postdoctoral fellows Ramon Alvarez-Puebla, Mathew Halls and Carlos Constantino, for their valuable input and comments on the manuscript.

Second, to all my friends and colleagues who have collaborated with me in the investigation of surface-enhanced vibrational spectroscopy, from whom, and with whom, I have learned a great deal. I wish to specifically acknowledge Dr A. Brolo and Dr M. Moskovits for their insightful comments and suggestions.

Third, I am indebted to the National Science and Engineering Research Council of Canada, without whose continuous financial support of my research in surface enhanced spectroscopy this book would not have been possible.

Finally, I am eternally grateful to my wife for her undivided love and constant encouragement of this project, and whose sacrificial dedication over four decades has continually served to inspire me.

Glossary

Definitions given are related to the content of this book. For extended acronyms or definitions see references 1 and 2–4, respectively. For a window into the on's terminology see Walker and Slack (5), and to avoid confusions in the world of optical constants, the excellent recollection by Holm (6) is recommended.

Absorbance (*A*). The logarithm to the base 10 of the ratio of the spectral radiant power of incident, essentially monochromatic, radiation to the radiant power of transmitted radiation: $A = -\log T$. In practice, absorbance is the logarithm to the base 10 of the ratio of the spectral radiant power of light transmitted through the reference sample to that of the light transmitted through the solution, both observed in identical cells. T is the (internal) transmittance. This definition supposes that all the incident light is either transmitted or absorbed, reflection or scattering being negligible.

Absorption of electromagnetic radiation. The transfer of energy from an electromagnetic field to matter. A process by which light is removed from the incident beam. This can include exciting electrons to higher energy states, transfer of light into heat or activation of various vibrational or rotational modes.

Absorptance. The fraction of light absorbed, equal to one minus the transmittance (T) plus reflectance (R).

Absorption band. This a region of the absorption spectrum in which the absorbance includes a maximum.

Absorption coefficient (decadic a or Napierian a). Absorbance divided by the optical pathlength: $a = A/l$. Physicists usually use natural logarithms. In this case, $\alpha = a \ln 10$, where a is the Napierian absorption coefficient. Since absorbance is a dimensionless quantity, the coherent SI unit for a and α is m^{-1}. Also cm^{-1} is often used.

Absorption cross-section (σ). Molecular entities contained in a unit volume of the absorbing medium along the light path. Operationally, it can be calculated as the absorption coefficient divided by the number of molecular entities contained in a unit volume of the absorbing medium along the light path: $\sigma = \alpha / N$.

Absorption spectrum. A two-dimensional plot of the absorbance or transmittance of a material with respect to wavelength or some function of the wavelength.

Angle of incidence. The angle at which the light beam strikes a surface. This angle is measured from the normal to the surface.

Anti-Stokes lines. These are Raman lines observed on the shorter wavelength side of the monochromatic radiation source. They arise from those Raman transitions in which the final vibration level is lower than the initial vibrational level.

Amphiphiles. Molecules with one part hydrophilic (water-loving) and the other part hydrophobic (water-hating). These are the most common monolayer-forming materials. The hydrophobic part is necessary to avoid the immersion of the molecule in the water subphase. The hydrophilic part is necessary to allow the spreading of the molecule on the water surface.

Analyte. In chemical analysis, the substance to be assessed is termed the analyte.

Attenuated total reflectance (ATR) (internal reflection spectroscopy). ATR is a reflectance sampling technique which is useful for analysis of liquids, polymer films and semi-solids. In ATR, infrared radiation impinges on a prism of infrared transparent material of high refractive index. Because of internal reflectance, the light reflects off the crystal surface at least once before leaving it. The infrared radiation sets up an evanescent wave which extends beyond the surface of the crystal into the sample that is in contact with the crystal.

Blinking. At the single molecule level, repeated cycles of fluorescent emission ('blinking') on a time-scale of several seconds are observed. This behavior would be unobservable in bulk studies.

Chemisorption. Metal–molecule interaction strongly alters the molecular electronic distribution owing to the formation of a chemical bond between molecule and the metal (surface complex), and consequently frequencies should be shifted.

Colloid. A heterogeneous system consisting of small (1–100 nm) particles suspended in a solution.

Electric susceptibility. For most common dielectric materials, the strength of the induced polarization P is proportional and parallel to the applied electric field E. Provided the field does not become extremely large and the medium is isotropic, $P = \varepsilon_0 \chi_e E$, where the constant χ_e is the *electric susceptibility* of the medium.

Electric displacement, D **($\mathrm{C\,m^{-2}}$).** For substances other than ferroelectric, the presence of an applied electric field, E, induces an electric polarization P, proportional to the magnitude of the applied field. For most common materials and weak fields, the response is linear and isotropic: $D = \varepsilon E$. The proportionality constant, ε is the *electric permittivity*, which in the general case is known as the *dielectric tensor*.

Dispersion. The variation of the index of refraction with frequency is called dispersion. The Kramers–Kronig relations allows one to calculate the light absorption properties of a medium when its dispersion is known.

Dye. An organic molecule with absorption bands in the visible spectral region.

Excimer. An excited dimer, dissociative in the ground state, resulting from the reaction of an excited molecule with a ground-state molecule of the same type.

Exciplex. An excited complex, dissociative in the ground state, resulting from the reaction of an excited molecule with a ground-state molecule of a different type.

Fermi energy. This is defined at absolute zero temperature. All orbitals of energy below the Fermi energy are occupied and all orbitals of higher energy are unoccupied. Notably, in the field of solid-state physics the chemical potential (temperatute dependent) is often called the *Fermi level*.

Fluorescence. Spontaneous emission of radiation (*luminescence*) from an excited molecular entity with the formation of a molecular entity of the same spin multiplicity.

Frank–Condon principle. Classically, the Frank–Condon principle is the approximation that an electronic transition is most likely to occur without changes in the position of the nuclei in the molecular entity and its environment. The resulting state is called the Frank–Condon state, and the transition involved, a vertical transition. The quantum mechanical formulation of this principle is that the *intensity* of a vibronic transition is proportional to the square of the overlap integral between the vibrational wavefunctions of the two states that are involved in the transition.

Full width at half-height or half-maximum (FWHH or FWHM). This is the width of the transmittance (absorbance or scattering) band measured at half the maximum transmittance (absorbance or scattering) value.

Langmuir film. Floating monomolecular film on the liquid subphase (usually water because its high surface tension).

Langmuir–Blodgett (LB) film. Film (monolayer or multilayer) fabricated transferring the Langmuir film from the liquid surface on to a solid substrate by the vertical movement of this solid substrate through the monolayer–air interface (like immersing a cookie in a mug of coffee). There are three types of LB films, called Z-type (transfer on the upstroke only), X-type (transfer on the down stroke only) and Y-type (transfer on the upstroke and down stroke).

Linewidth. The linewidth of the *particle-plasmon resonance* is controlled by lifetime broadening due to various decay processes. Part of this lifetime broadening results from nonradiative decay of the particle plasmon into electron–hole excitations in the metal; if the excitations occur within the conduction (s–p) band, the decay process is termed *intraband damping*. If the excitations are between d bands and the conduction band, it is called *interband damping*.

Near-field. The near-field can be defined as the extension outside a given material of the field existing inside this material. In most cases, the amplitude of the near-field decays very rapidly along the direction perpendicular to the interface, giving rise to the so-called evanescent wave character of the near-field. The most relevant to SEVS are surface near-fields that can only be produced by applying an external excitation (photon excitation).

Organic semiconductors. From the band theory point of view, there is not much difference between organic and inorganic semiconductors. In a solid, the density is so high that the interatomic spacing becomes very small. The interaction of the atoms causes each of the original atomic orbital to split into N components; since N is a extremely large number, the spacing between the energy levels becomes negligibe and the individual levels coalesce into an energy band. The valence levels produce a valence band and the allowed higher levels produce a conduction band. These two bands are separated by an energy gap or forbidden zone. In a metal, the uppermost energy band is partially filled or a filled band overlaps an empty band, then there are some electrons free to move in a field, resulting high conductivities. In insulators, the valence band is full, the conduction band is empty and the energy gap is of several electronsvolts, and no electrons are able to carry current. A semiconductor stands between these two extremes the energy gap is around 1 eV (Si 0.7, Ge 1.2, phthalocyanines 1.68 eV). So, if the system is properly excited, electrons will promote to the conduction band being able to carry current.

Phonon. The quantum of energy of an elastic wave in a solid. A quantum of sound. The thermal average number of phonons in an elastic wave of frequency ω is given by the Planck distribution function, just as for photons.

Plasmons (or surface excitation of electron–hole pairs). These are simply the quanta of the oscillations of the surface charges produced by external electric field. Plasmon modes can be sustained in thin films, called surface plasmons (SPs), and in nanoparticles, called localized SPs or particle plasmons (PPs). Surface plasmons on a plane surface are non-radiative electromagnetic modes. The origin of the non-radiative nature of SPs is that the interaction between light and SPs cannot simultaneously satisfy energy and momentum conservation. This restriction can be circumvented by relaxing the momentum conservation requirement by roughening or corrugating the metal surface. A second method is to increase the effective wavevector (and hence momentum) of the light by using a prism coupling technique.

Photo-excitation. The production of an excited state by the absorption of ultraviolet, visible or infrared radiation.

Polarizability. When an electric field is applied to an individual atom or molecule, the electron distribution is modified and the molecular geometry is distorted. Atoms and molecules respond to electric fields by acquiring an electric dipole moment (in addition to the one they may already possess) as the centroids of positive and negative charge are displaced. The polarizability, α, is the constant of proportionality between the induced dipole moment, μ, and the strength of the electric field, $E : \mu = \alpha E$. If the applied field is very strong, the induced dipole also depends on $E^2$and higher powers; the coefficients of the higher power of E are known as hyperpolarizabilities. The total polarizability of a system can be divided into several contributions. The atomic polarizability is the contribution of the geometric distortion. It is usually significantly smaller than the electronic polarizability, which is the contribution from the displacement of the electrons.

Raman effect. The inelastic scattering, i.e. scattering with change in the frequency of the incident radiation passing through a substance, is called Raman scattering. In the spectrum of the scattered radiation, the new frequencies are termed Raman lines, or bands, and collectively are said to constitute a Raman spectrum. Raman bands observed at frequencies lower than the exciting laser frequency are referred to as *Stokes bands*, and those at frequencies greater than the incident laser frequency as *anti-Stokes bands*.

Rayleigh scattering. This is the incoherent and elastic scattering of light by particles much smaller than the wavelength of the incident radiation. The scattering intensity is inversely proportional to the fourth power of the incident wavelength, and about 1 part in 10^3 of the incident radiation undergoes Rayleigh scattering.

Reflection absorption infrared spectroscopy (RAIRS). This technique probes the interface region above a metal surface by measuring the absorption of a specularly reflected infrared beam, incident at glancing angles, as a function of wavenumber.

Relative permittivity. With the definition of electric displacement, $D = \varepsilon_0 E + P$ and $P = \varepsilon_0 \chi_e E$, the electric permittivity is $\varepsilon = \varepsilon_0 (1 + \chi_e)$. Materials are commonly classified according to their *relative permittivity* or dielectric constant, a dimensionless quantity ε_r, defined as the ratio $\varepsilon_r = \varepsilon / \varepsilon_0$.

Physisorption. Metal–molecule interaction due to Van der Waals type force, and does not result in a substantial change in the vibrational energy levels, i.e. vibrational frequencies will be observed unshifted from their values in the absence of the metal surface.

Signal-to-noise ratio (SNR). This SNR is used to measure the quality of a spectrum. The ratio of the signal in a spectrum, usually measured as the intensity of an absorbance band, to the noise measured at a nearby point in the baseline determines this SNR value.

Wavenumber. The units of wavenumbers are cm^{-1}, and are most commonly used as the X-axis unit in infrared and Raman spectra. It indicates how many waves can fit in 1 cm.

The surface-enhanced family:

SERS. Surface-enhanced Raman scattering.
SERRS. Surface-enhanced resonant Raman scattering.
FT-SERS. Fourier transform surface-enhanced Raman scattering.
NIR-SERS. Near-infrared surface-enhanced Raman scattering.
TERS. Tip-enhanced Raman scattering.
SEIRA. Surface-enhanced infrared absorption.
SEIRRA. Surface-enhanced infrared reflection–absorption.
SEF. Surface-enhanced fluorescence (also MEF: Metal-enhanced fluorescence).
SES. Surface-enhanced spectroscopy.
SMS. Single molecule spectroscopy.
SMD. Single molecule detection.
SESHG. Surface-enhanced second harmonic generation.
SEHRS. Surface-enhanced hyper-Raman spectroscopy.

REFERENCES

[1] D.A.W. Wendisch (1990). Acronyms and Abbreviations in Molecular Spectroscopy. Berlin: Springer-Verlag.

[2] J.W. Verhoeven (1996). International Union of Pure And Applied Chemistry. Organic Chemistry Division. Commission on Photochemistry. Glossary of Terms Used in Photochemistry. (IUPAC Recommendations 1996). Pure & Appl. Chem. **Vol. 68.** 2223–86.

[3] S.P. Parker (Ed).(1988). Solid-State Physics Source Book. New York: McGraw-Hill Book Company.

[4] R.G., Lerner G.L. Trigg (Ed).(1991). Encyclopedia of Physics. New York: VCH Publishers, Inc.

[5] C.T., Walker G.A. Slack. (1970). Who named the -ONs? American Journal of Physics **38.** 1380–89.

[6] R.T. Holm (1991). in E.D. Palik (Ed.), Handbook of Optical Constants of Solids II (Pages 21–55). New York: Academic Press.

1

Theory of Molecular Vibrations. The Origin of Infrared and Raman Spectra

1.1 ELECTRONIC, VIBRATIONAL, ROTATIONAL AND TRANSLATIONAL ENERGY

In classical mechanics, a molecule can be seen as a collection of M nuclei and N electrons. Therefore, the system of $M + N$ particles has $3(N + M)$ degrees of freedom to describe its motions. First, one can fix in space the location of the heavy nuclei (fixed nuclei approximation). The symmetry of this spatial distribution of nuclei can be associated with a 'molecular point group', which is a symmetry group corresponding to a fixed point [the center of mass (CM)]. The $3N$ degrees of freedom describe the motion of the electrons around the frozen frame, and the corresponding energy of motion is the electronic energy E_e. We can regroup the nuclei and electrons into $3M$ effective atoms, and fix the origin of the system of coordinates in the CM of the molecule. The motion of this point in space is described by three degrees of freedom, and gives the translational energy of the molecule that is directly related to thermal energy. According to the equipartition principle, the energy is $3/2kT$, where k is the Boltzmann constant. For 1 mol of molecules, we multiply by Avogadro's number, N_A, and k is simply replaced by $N_A k = R$, the gas constant,

Surface-Enhanced Vibrational Spectroscopy R. Aroca

and the thermal energy per mole is $3/2RT$. For the fixed molecule at the CM there are $3M - 3$ degrees of freedom. The fixed molecule can rotate, and to describe the rotation of a nonlinear molecule we need three degrees of freedom (two for a linear molecule). Therefore, we can eliminate six of the $3M$ coordinates and we are left with $3M - 6$ (or $3M - 5$) vibrational degrees of freedom to describe the motions of the nuclei (effective atoms), and the total energy of the molecule has been partitioned into electronic, vibrational, rotational and translational (thermal) [1–3]:

$$E_{\text{molecular}} = E_{\text{electronic}} + E_{\text{vibrational}} + E_{\text{rotational}} + E_{\text{translational}}$$

1.1.1 Electronic Structure of Molecules

The origin of electronic, vibrational and rotational spectroscopy is in the quantization of these energies, and we shall briefly refresh the quantum mechanical treatment of molecules [4,5]. In spectroscopy, the word molecule refers to a stable system of nuclei and electrons. When the total number of electrons differs from that of the positive charges, the system is said to be a molecular ion. When the number of electrons is odd, the system is called a free radical (a free radical is defined as a system with a nonzero spin). Nuclei and electrons have well-defined mass, charge and spin. Since molecules are made of nuclei and electrons, molecules have well defined mechanical (mass), electrical (charge) and magnetic (spin) properties. In particular, the ratio of the mass of the proton to the mass of the electron is 1836. Therefore, the mass of the nuclei is at least 1836 times larger than the mass of the electrons. This fact allows for the separate treatment of the motion of the electrons (electronic spectrum) from that of the nuclei (vibrational spectrum) [6].

The total molecular Hamiltonian, $\hat{H}_{\text{MOL}}$, describes a molecule isolated in space, that is, no external field is acting upon the molecule. The external potential V_{ext} equals zero. Further, the total molecular Hamiltonian is written solely in terms of the spatial coordinates, i.e. the spin variables are not included in the Hamiltonian. In the spinless molecular Hamiltonian, two terms can be distinguished:

$$\hat{H}_{\text{MOL}} = T + V \tag{1.1}$$

Where T is the kinetic energy operator of all M nuclei and N electrons of the system:

$$T = -\frac{\hbar^2}{2m}\sum_{i=1}^{N}\Delta_i - \frac{\hbar^2}{2}\sum_{\alpha=1}^{M}\frac{1}{M_\alpha}\Delta_\alpha = T_e + T_n \tag{1.2}$$

where

$$\nabla \cdot \nabla = \nabla^2 = \Delta = \frac{\partial^2}{\partial x^2} + \frac{\partial^2}{\partial y^2} + \frac{\partial^2}{\partial z^2}$$

The subscript i represents the number of electrons, α is the number of nuclei, M_α is the mass of the nucleus α, T_e is the electronic kinetic energy operator and T_n is the nuclear kinetic energy operator. The terms of the potential energy operator V_T can be classified, as in the case of atoms, into two parts: electrostatic interactions and interactions between momenta:

$$V_T = -\sum_{i}^{N}\sum_{\alpha}^{M}\frac{z_\alpha e^2}{R_{i\alpha}} + \frac{1}{2}\sum_{i\neq j}\frac{e^2}{r_{ij}} + \frac{1}{2}\sum_{\alpha\neq\beta}\frac{z_\alpha z_\beta e^2}{R_{\alpha\beta}} + V'. \tag{1.3}$$

where r_{ij} is the distance between two electrons i and j, $R_{i\alpha}$ is the distance between the electron i and the nucleus α and $R_{\alpha\beta}$ is the distance between two nuclei α and β. The first term is the electron–nuclear attraction, the second term is the electron–electron repulsion and the third term is the nuclear–nuclear repulsion. V' includes the interactions between the spin angular momenta of nuclei and electrons and the orbital angular momenta of electrons:

$$V' = V_{(\text{spin-orbit})} + V_{(\text{spin-spin})} \tag{1.4}$$

In what follows, only the electrostatic interactions will be taken into account; the interaction between momenta may be considered as a perturbation. Hence the potential energy operator (1.3) can be rewritten as

$$V = V_{ee} + V_{en} + V_{nn} \tag{1.5}$$

where V' has been neglected (later on it can be included as a perturbation to the basic electrostatic problem). The electrostatic potential energy operator (1.5) is a function of the distances between nuclei and electrons only, and a separation of variables can be carried out on the stationary Schrödinger equation. This means that the three degrees of freedom corresponding to the CM of the system can be separated. Therefore, the Schrödinger equation is a differential equation of $3(N+M)-3$ variables. Such an equation cannot be solved for most of molecular systems. Under these circumstances, a variety of approximate approaches are used. All these approximate methods have, however, a common starting point: the separation of nuclear and electronic motions, which is known as the Born–Oppenheimer or adiabatic approximation [3,4,7]. The foundation for the approach is the assumption of a large energy splitting between the electronic states. Notably, for molecules adsorbed on metal surfaces the use of the approximation may come into question [8].

1.2 SEPARATION OF NUCLEAR AND ELECTRONIC MOTIONS

The eigenfunction $\psi(r,R)$, with r being electron coordinates and R nuclear coordinates, in the stationary Schrödinger equation is approximated by a product:

$$\psi(r,R) = \Phi(r,R)\chi(R). \tag{1.6}$$

The function $\Phi(r,R)$ depends on R only in a parametric fashion and is known as the electronic wavefunction, and satisfies the completeness relation

$$\langle\Phi(r,R)|\Phi(r,R)\rangle = 1 \tag{1.7}$$

where the integration is only over electronic coordinates. The function $\chi(R)$ is known as the *nuclear wavefunction* and satisfies the condition

$$\langle\chi(R)|\chi(R)\rangle = 1. \tag{1.8}$$

Here the integration takes place over nuclear coordinates only. On the basis of the variational principle, it can be shown that the function $\Phi(r,R)$

is determined by

$$H_e\Phi = (T_e + V)\,\Phi = E_e\,(R)\,\Phi \tag{1.9}$$

where V is the electrostatic potential operator (1.5), $E_e(R)$ are the eigenvalues of the electronic equation and are functions of the nuclear coordinates in a parametric form.

The function $\chi(R)$ is the solution of the equation

$$H_n\chi(R) = [T_n + E_e(R)]\chi(R) = E\chi(R) \tag{1.10}$$

Where E is the total energy of the system. Equation (1.10) is known as the *nuclear equation*. Let us assume that the electronic equation has been solved for fixed values of nuclear coordinates R_0. Each eigenvalue $E_e(R_0)$ depends on the nuclear coordinates as parameters. Let us take the lowest energy eigenvalue $E_e^0(R_0)$ and study its dependence with variations in nuclear coordinates. A plot of the $E_{\mathrm{e}}^0(R_0)$ values against the internuclear distance in a diatomic case gives rise to well-known potential energy curves. For the case of more than two nuclei a potential energy surface (or hyper surface) is obtained. It is usually the analytical form of this dependence that is included in Equation (1.10) as a potential energy operator $E_e(R)$. For M nuclei there exist $3M$ nuclear coordinates. Assuming that the center of mass is entirely determined by the nuclei, the total number of nuclear coordinates is reduced to $3M - 3$. Of these $3M - 3$ nuclear coordinates, only three are needed to describe the rotation of a nonlinear system in a frame of reference mounted on the molecule with its origin at the center of mass. The other $3M - 6$ nuclear coordinates of a nonlinear molecule describe the vibrational motion of the nuclei within the molecule. For a linear molecule, these numbers are 2 for rotational coordinates and $3M - 5$ for vibrational coordinates. It can be seen that the nuclear Equation (1.10) may be subjected to a 'second' Born–Oppenheimer approximation that will allow one to separate a vibrational equation with eigenvalues E_v and a rotational equation with eigenvalues E_R. In a first approximation, then, the quantized energy of a fixed molecule can be represented as the sum of three parts: the electronic, the vibrational and the rotational energies:

$$E_{\mathrm{TOTAL}} = E_e + E_v + E_R. \tag{1.11}$$

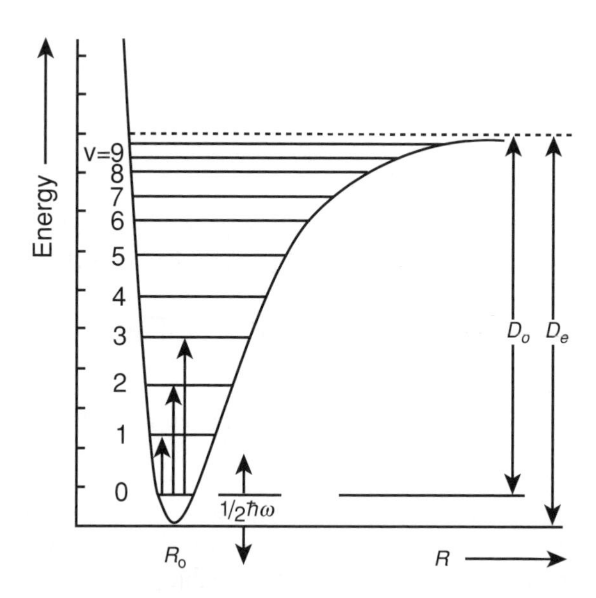

Figure 1.1 Potential energy curve of a diatomic molecule in the ground electronic state with vibrational energy levels. R is the internuclear distance. The electronic energy difference D_e is greater than D_0, the dissociation energy or heat of dissociation

1.2.1 Example. The Potential Energy Function of Diatomic Molecules

A diatomic molecule can exist in the ground electronic state and also in a series of excited electronic states. Each electronic state is determined by an electronic wavefunction $\psi_e(r, R)$ and an electronic energy $E_e(R)$. The exact form or analytical expression of the function $E_e(R)$ for each electronic state of the molecule can be obtained by solving the electronic Equation (1.1) for different values of the internuclear distance R. A typical potential function and vibrational energy levels in the ground state of a diatomic molecule are shown in Figure 1.1. In molecular spectroscopy and statistical thermodynamics, it is common to set the origin equal to the energy minimum of the ground electronic state, i.e. $E_e^{(0)}(R_e) = 0$. This convention has been applied in Figure 1.1.

Once the potential energy curve has been found, the main characteristics of the electronic state are defined by:

1. The electronic energy value at the minimum of the potential energy curve; $E_e^{(0)}$.
2. The equilibrium internuclear distance, R_e, which is the internuclear distance at the minimum of the potential energy curve.

3. The potential energy of dissociation, D_e, which is the difference between the dissociation limit $E_e^{(\infty)}$ and the minimal value of the electronic energy $E_e^{(0)}$: $D_e = E_e^{(\infty)} - E_e^{(0)}$.
4. The second derivative of the electronic energy with respect to the internuclear distance; this quantity is known as *force constant* or potential constant:

$$k_e = \left[\frac{d^2 E_e^{(R)}}{dR^2}\right]_e.$$

Different electronic states are characterized by different values for E_e, R_e, D_e, k_e and ω_e (the harmonic vibrational frequency). Typical values for diatomic molecules are given in Table 1.1.

1.3 VIBRATIONS IN POLYATOMIC MOLECULES

The same semiclassical treatment for the vibrational motion of the nuclei on the potential energy surface provided by the electronic energy function can be extended to polyatomic molecules [1,3,7,9,10]. For a system of N

Table 1.1 Observed spectroscopic constants and calculated potential constant for a selection of diatomic molecules

Molecule	R_e/Å	k_e/mdyn Å^{-1}	ω_e/cm^{-1}	D_0/eV
H_2	0.741	5.755	4402.7	4.48
C_2	1.242	12.160	1854.7	6.24
N_2	1.097	22.940	2358.1	9.76
O_2	1.207	11.768	1580.4	5.12
F_2	1.417	4.451	891.9	1.60
Cl_2	1.987	3.227	559.7	2.48
Br_2	2.281	2.461	323.3	1.97
I_2	2.665	1.720	214.5	1.54
LiH	1.595	1.026	1405.7	2.43
BH	1.232	3.048	2366.9	3.46
CH	1.120	4.478	2859.1	3.45
NH	1.047	5.41	3125.5	3.21
OH	0.970	7.793	3735.2	4.39
FH	0.917	9.651	4137.3	5.86
ClH	1.274	5.163	2991.0	4.43
BrH	1.406	4.166	2650.0	3.76
IH	1.609	3.140	2308.6	3.05
BO	1.205	13.658	1885.4	4.60
CO	1.128	19.019	2169.8	11.10
NO	1.150	15.948	1904.0	6.50

nuclei with nonlinear geometry there are $3N-6$ vibrational degrees of freedom and for a linear equilibrium geometry there are $3N-5$ vibrational degrees of freedom. Working within the model of the harmonic oscillator [3,7], the potential energy can be written as

$$V=\frac{1}{2}\sum_{i,j=1}^{3N-6}k_{ij}q_iq_j \tag{1.12}$$

where

$$k_{ij}=\left(\frac{\partial^2 V}{\partial q_i\partial q_j}\right)_{q_i=q_j=0} q_iq_j \tag{1.13}$$

or, in matrix form,

$$2V=\{q\}\,U_q\,\|q\|\,. \tag{1.14}$$

In the same way the kinetic energy is given by

$$2T=\{\dot{q}\}\,T_q\,\|\dot{q}\| \tag{1.15}$$

where the 'dot' notation represents differentiation with respect to time.

Replacing Equation (1.14) and (1.15) in the Lagrange equation:

$$\frac{\mathrm{d}}{\mathrm{d}t}\frac{\partial L}{\partial \dot{q}_i}-\frac{\partial L}{\partial q_i}=0 \qquad (L=T_q-U_q)$$

a system of $n=3N-6$ linear differential equations is obtained:

$$\sum_{j=1}^{n}(t_{ij}\ddot{q}_j+k_{ij}q_j)=0. \tag{1.16}$$

Considering a solution of the form

$$q_j=l_j\cos(\omega t+\delta)=l_j\cos(\sqrt{\lambda}t+\delta) \tag{1.17}$$

Equation (1.16) transforms to

$$\sum_{j=1}^{n}(k_{ij} - \lambda t_{ij})l_j = 0 \qquad (1.18)$$

or, in matrix form,

$$(U_q - \lambda T_q)L = 0. \qquad (1.19)$$

The problem then reduces to finding the eigenvalues and eigenvectors of the secular equation

$$\det|U_q - \lambda T_q| = 0. \qquad (1.20)$$

The molecular vibrational problem of polyatomic molecules is reduced to solving the secular Equation (1.20). This equation, however, is not convenient for practical computations. Thus, multiplying by T_q^{-1} from the left, $\det|T_q^{-1}U_q - \lambda I| = 0$, where I is the unit matrix. It has been assumed that the coordinates q form an independent set of coordinates, otherwise, T_q^{-1} would not exist. The last equation in matrix from is written as

$$T_q^{-1}U_qL = L\Lambda \qquad (1.21)$$

and introducing Wilson's notation [4], $T_q^{-1} = G$ and $U_q = F$, Equation (1.21) finally gives

$$GFL = L\Lambda. \qquad (1.22)$$

Equation (1.22) is usually known as the G-Wilson method for molecular vibration. G is thus the inverse of the kinetic energy matrix. Practical problems are related to the finding of G- and F-matrix elements.

In quantum mechanics, the harmonic approximation for a nonlinear molecule gives a discrete spectrum of energy values:

$$E_0 = \sum_{i=1}^{3N-6}\hbar\omega_i\left(v_i + \frac{1}{2}\right) \quad v_i = 0, 1, 2, \ldots \qquad (1.23)$$

where v_i is the vibrational quantum number and ω_i is the harmonic vibrational frequency. Potential energy surfaces for polyatomic molecules can be obtained using *ab initio* Hartree–Fock (HF) and density functional theory (DFT) methods that are now common analytical tools for infrared and Raman spectral computations. Thereby the $3N-6$ or $3N-5$ normal modes of the harmonic approximation can be found. The symmetry of the potential function will allow for the reduction in size of the matrix (1.22) to a group of smaller matrices, one for each irreducible representation of the molecular point group. The methods of group theory will permit the calculation of the number of normal modes in each of the symmetry species and the extraction of their infrared or Raman activity.

The vibrational problem, or finding the infrared and Raman frequencies and intensities, is currently solved directly using quantum chemistry, and we will illustrate this computational approach using Gaussian 98. The detailed example at the end of this chapter was chosen to illustrate the applications to surface-enhanced vibrational problems.

1.4 EQUILIBRIUM PROPERTIES. DIPOLE MOMENT AND POLARIZABILITY

The interpretation of the observed infrared and Raman spectra using the basic models of the rigid rotator and harmonic oscillator are explained in Herzberg's book (Chapter III, p. 66) [2]. This approximation is the basis for the widespread application of vibrational spectroscopy as a tool for the detection, identification and characterization of molecules.

Two molecular properties that are defined by the charge distribution at the equilibrium geometry of the electronic state will change with variations in the internuclear distance (or any of the vibrational degrees of freedom in a polyatomic molecule): the dipole moment μ and the molecular polarizability α. The dipole moment is a vector, $\mu = \mu_x + \mu_y + \mu_z$, and for each of the components we can write a series expansion about the equilibrium geometry:

$$\mu = \mu_0 + \left(\frac{\partial \mu}{\partial q}\right)_0 q + \frac{1}{2}\left(\frac{\partial^2 \mu}{\partial q^2}\right)_0 q^2 + \ldots \tag{1.24}$$

where μ_0 represents the equilibrium value of the dipole moment. The displacement q has the form $q(t) = q_0 \cos(\omega_0 t)$. It will be seen that the infrared spectrum of fundamental vibrational frequencies is determined by the first partial derivative$(\partial\mu/\partial q)_0$ in the series [11,12]. Since there are three components for each vibration, each vibrational frequency has up to three chances to be seen in the infrared spectrum. In other words, for a vibrational transition to be allowed in the infrared spectrum, it is necessary that at least one of these three components be different from zero. Notably the first term, the permanent dipole moment, μ_0, will play no role in the probability of seeing a fundamental vibration in the infrared spectrum. The polarizability is a tensor, a response function that represents the volume and shape of the molecular electronic cloud [13]:

$$\alpha = \begin{pmatrix} \alpha_{xx} & \alpha_{xy} & \alpha_{xz} \\ \alpha_{yx} & \alpha_{yy} & \alpha_{yz} \\ \alpha_{zx} & \alpha_{zy} & \alpha_{zz} \end{pmatrix}. \tag{1.25}$$

For spectroscopic applications, the tensor is considered to be symmetric, reducing the total number of unknowns for each tensor to six. As was seen for the dipole moment, each of the components of the polarizability tensor can be written as a series expansion about the equilibrium geometry:

$$\alpha = \alpha_0 + \left(\frac{\partial \alpha}{\partial q}\right)_0 q + \frac{1}{2}\left(\frac{\partial^2 \alpha}{\partial q^2}\right)_0 q^2 + \ldots \tag{1.26}$$

where α_0 is the equilibrium value of the polarizability tensor element, and q represents the deviation from equilibrium. The first derivative, $\alpha' = (\partial\alpha/\partial q)_0$, is responsible for determining the observation of vibrational fundamentals in the Raman spectrum [13]. Since polarizability is a response function of the molecule to an external electric field, the polarizability and polarizability derivatives are tensors of the second rank, i.e. for a symmetric tensor each vibration has six chances to be observed in the Raman spectrum. In other words, for a vibrational transition to be allowed in the Raman spectrum, it is necessary that at least one of the six components of the derivative tensor be different from zero. The polarizability derivative tensor (the Raman tensor) is shown in Equation (1.27), where the first partial derivative is represented by α'_{ij}. The α' tensor has certain important properties: it is symmetric and its trace is invariant.

$$\alpha' = \begin{pmatrix} \alpha'_{xx} & \alpha'_{xy} & \alpha'_{xz} \\ \alpha'_{yx} & \alpha'_{yy} & \alpha'_{yz} \\ \alpha'_{zx} & \alpha'_{zy} & \alpha'_{zz} \end{pmatrix}. \tag{1.27}$$

1.5 FUNDAMENTAL VIBRATIONAL TRANSITIONS IN THE INFRARED AND RAMAN REGIONS

The description of the energy states and equilibrium properties of the molecule given above has prepared us for the final step in explaining infrared and Raman spectra: the interaction of the molecule with electromagnetic radiation [13]. The interaction of the electric vector of the electromagnetic radiation with the molecule will give rise to infrared absorption and inelastic scattering (Raman) spectra [14]. The simplest description of the electric field of light is that of plane harmonic waves, which can be written as

$$E(r,t) = E_0 \exp[i(kr - \omega t)] \tag{1.28}$$

where the vector E is perpendicular to k, the propagation direction. The direction of E in space determines its polarization and, thereby, for a wave traveling along z, the light polarization can be either E_x or E_y (light polarization is discussed in Chapter 2). We have a quantum object, the molecule, interacting with the radiation field, a plane wave, as described in classical electromagnetic theory. Thereby, the description of the process is semi-classical, and the interaction is known as the semi-classical theory of quantum transitions. The coupling operator between the quantum molecule and the radiation field is given by $H' = -p \cdot E$, where p is the dipole moment and E is the electric field vector. For infrared p is given by Equation (1.24) and for Raman by $p = \alpha E$. The probability for the absorption (or emission) of electromagnetic radiation per unit time is proportional to the square of the transition dipole moment matrix element along the direction of the light polarization: $|\langle \Psi_v | p \cdot E | \Psi_{v'} \rangle|$. For molecules in the gas phase with a random orientation, and where the average of the square of the angular part is one, the discussion can proceed with $|\langle \Psi_v | p | \Psi_{v'} \rangle|$.

The Raman effect can be explained in terms of the induced dipole moment, $p = \alpha E$, and using Equation (1.26) for the polarizability:

$$p = \alpha_0 E + \left(\frac{\partial \alpha}{\partial q}\right)_0 qE + \frac{1}{2}\left(\frac{\partial^2 \alpha}{\partial q^2}\right)_0 q^2 E + \ldots. \tag{1.29}$$

However, since $E = E_0 \cos(\omega t)$, and neglecting the second derivative,

$$p = \alpha_0 E_0 \cos(\omega t) + \left(\frac{\partial \alpha}{\partial q}\right)_0 q_0 \cos(\omega_0 t) E_0 \cos(\omega t) \tag{1.30}$$

where ω_0 is the natural vibrational frequency of the molecule and ω is the frequency of the radiation field. Using the trigonometric identity $\cos a \cdot \cos b = \frac{1}{2}[\cos(a+b) + \cos(a-b)]$, the induced dipole expression is

$$p = \alpha_0 E_0 \cos(\omega_0 t) + \alpha' q_0 E_0 \cos(\omega_0 - \omega)t + \alpha' q_0 E_0 \cos(\omega_0 + \omega)t. \tag{1.31}$$

The first term will account for the elastic Rayleigh scattering, the second for the Stokes Raman scattering and the third for the anti-Stokes Raman scattering.

For infrared absorption, neglecting the 'electrical anharmonicity' (second derivatives) in Equation (1.24), the transition between two vibrational states is $|\langle \Psi_v | \mu' | \Psi_{v'} \rangle|$, with $\mu' = (\partial \mu / \partial q)_0$. Dipole moment derivatives μ' form a three-dimensional vector, i.e. $\mu' = \mu'_x + \mu'_y + \mu'_z$.

The computation of the transition dipole moment matrix element will answer the question of whether a particular vibrational transition would be allowed or forbidden in the infrared or Raman spectrum. The results are known as the selection rules for infrared and Raman spectra. Ultimately, it should be remembered that for molecular systems other than gases (crystals, organized films, adsorbed molecules and others), the final observation of a particular vibrational transition in the infrared or Raman spectrum also depends on the direction of the incident radiation field.

1.6 SYMMETRY OF NORMAL MODES AND VIBRATIONAL STATES

The first task in the study of the vibrational spectrum of a given molecule should be the finding of the symmetry point group to which the equilibrium molecular geometry belongs. Group theory is discussed in specialized undergraduate textbooks, and we will review here only the basic elements relevant to vibrational spectroscopy [12,15,16]. Groups are a set of operations that satisfy the following four conditions: (i) one of the operations is the identity operation; (ii) each operation in the group has an inverse; (iii) the members of the group fulfill the associative law; and (iv) the product of two members of the group is also a member of the group. The symmetry operations that form the point groups transform the molecule into self-coincidence. Rotations are symmetry operations

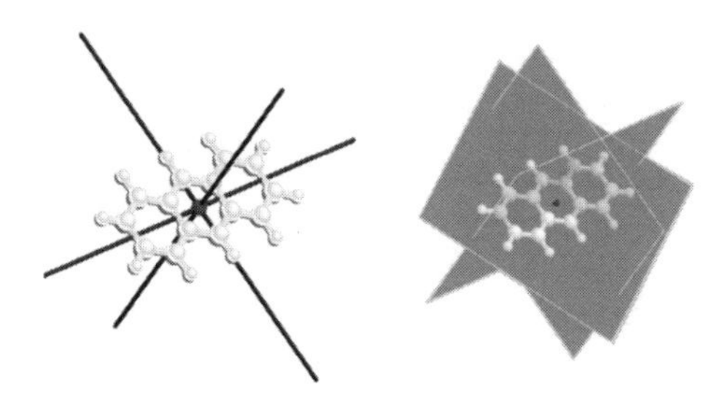

Figure 1.2 Axes of rotation and planes of symmetry in the anthracene molecule

that are called *proper* because they do not change the chirality of the molecule. Rotation–reflection operations are called *improper*, because they are not physically feasible and change the chirality. To carry out the operation of symmetry a *symmetry element* is necessary. This could be a point, a line or a plane. Therefore, symmetry operations are associated with symmetry elements. There are only four symmetry elements, the n-fold axis of rotation C_n, the n-fold rotation-reflection axis S_n, the plane σ and the center of inversion i. Finding them in a molecule allows the assignment of the molecule to one of the 32 point groups. Anthracene is a molecule belonging to the D_{2h} point group where the C_n, σ and i elements of symmetry can be found, and is used here for illustration.

Figure 1.2 (left) illustrates the two axes of rotations, C_2, found in the molecule. In Figure 1.2 (right), the three planes of symmetry and the center of symmetry are highlighted. Each group has a finite number of symmetry operations and, with them, the group multiplication table can be generated. Since symmetry operations are transformations of coordinates, each of them can be represented by a three-dimensional matrix. For a molecule a reducible matrix representation Γ can be constructed that contains a number of irreducible representations Γ_i. The trace of the matrix in the irreducible representation is called the *character* of the irreducible representation and is denoted with the Greek letter $\chi(R)$, where R represents the symmetry operation [15]. Point group character tables are given in almost all spectroscopy textbooks [16]. Every group contains the identity operation E. The sum of the squares of the characters under E gives the order of the group. The character table for the D_{2h} group of anthracene is given in Table 1.2.

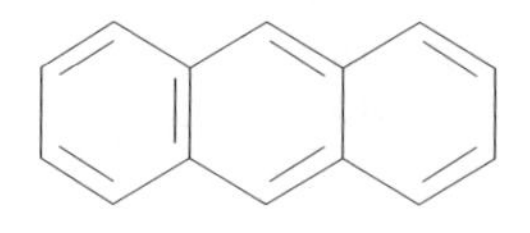

Table 1.2 Character table for the D_{2h} point group of anthracene

D_{2h}	E	$C_2(z)$	$C_2(y)$	$C_2(x)$	i	$\sigma(xy)$	$\sigma(xz)$	$\sigma(yz)$	μ'	A'
A_g	+1	+1	+1	+1	+1	+1	+1	+1		$\alpha'_{xx}, \alpha'_{yy}, \alpha'_{zz}$
B_{1g}	+1	+1	−1	−1	+1	+1	−1	−1		α'_{xy}
B_{2g}	+1	−1	+1	−1	+1	−1	+1	−1		α'_{xz}
B_{3g}	+1	−1	−1	+1	+1	−1	−1	+1		α'_{yz}
A_u	+1	+1	+1	+1	−1	−1	−1	−1		
B_{1u}	+1	+1	−1	−1	−1	−1	+1	+1	μ'_z	
B_{2u}	+1	−1	+1	−1	−1	+1	−1	+1	μ'_y	
B_{3u}	+1	−1	−1	+1	−1	+1	+1	−1	μ'_x	

The sum of the squares under E is equal to 8, the order of the group. A is a one-dimensional representation symmetric with respect to rotation about the principal axis. B is a one-dimensional representation anti-symmetric with respect to rotation about the principal axis. g (*gerade*) is symmetric with respect to the inversion centre and u (*ungerade*) is antisymmetric with respect to the inversion centre. The subscripts 1 (symmetric), 2 and 3 (antisymmetric) are used solely with A and B.

After the point group has been identified, we proceed to the assignment of the fundamental vibrational frequencies (normal modes) to the irreducible representations of the group. For the molecule at hand, anthracene, with 24 atoms, we have $(3 \times 24) - 6 = 66$ normal modes. Since there are three internal cycles in the molecule, the total number of stretching vibrations is $24 - 1 + 3 = 26$, which is also the total number of chemical bonds in the molecule. The 26 stretching modes contain 10 high-frequency C–H stretchings and 16 C–C ring stretching modes. The other 40 modes are deformation of plane angles (angle between three atoms) and dihedral angles (angle between two planes). All of these normal modes expressed in terms of generalized coordinates (Lagrange's formalism) are set up using $3N$ ($N =$ number of atoms) Cartesian displacement coordinates. During the symmetry operation, a number of atoms are shifted while a few remain unshifted (n_R). Accordingly, only unshifted atoms can contribute to the character, and a new character X_{vib} must be constructed to be added to the given character table. The calculation of this character is different for proper and improper rotations, and eliminating from the outset the pure translations and rotations to leave $3N - 6$ (or $3N - 5$) vibrations, the expressions are (note that reflections are improper rotations):

$$\chi_{\text{vib}}\left(C_n^k\right) = (n_R - 2)(1 + 2\cos\theta) \tag{1.32a}$$

Table 1.3 Contributions of each operation to character

Proper rotation		Improper rotation	
C_n^k	$+1+2\cos\theta$	S_n^k	$-1+2\cos\theta$
$C_1^1 = E$	$+3$	$S_1^1 = S_1^1 = \sigma$	$+1$
C_2^1	-1	$S_2^1 = i$	-3
C_3^1, C_3^2	0	S_3^1, S_3^5	-2
C_4^1, C_4^3	$+1$	S_4^1, S_4^3	-1
C_5^1, C_5^4	$+\tau$	S_5^1, S_5^9	$\tau-2$
C_5^2, C_5^3	$1-\tau$	S_5^3, S_5^7	$-1-\tau$
C_6^1, C_6^5	$+2$	S_6^1, S_6^5	0
C_7^1, C_7^6	$+1+2\cos\frac{2\pi}{7}$	S_7^1, S_7^{13}	$-1+\cos\frac{2\pi}{7}$
C_7^2, C_7^5	$+1+2\cos\frac{4\pi}{7}$	S_7^3, S_7^{11}	$-1+2\cos\frac{4\pi}{7}$
C_7^3, C_7^4	$+1+2\cos\frac{6\pi}{7}$	S_7^5, S_7^9	$-1+2\cos\frac{6\pi}{7}$
C_8^1, C_8^7	$1+\sqrt{2}$	S_8^1, S_8^7	$-1+\sqrt{2}$
C_8^3, C_8^5	$1-\sqrt{2}$	S_8^3, S_8^5	$-1-\sqrt{2}$
C_{10}^1, C_{10}^9	$1+\tau$	S_{10}^1, S_{10}^9	$-1+\tau$
C_{10}^3, C_{10}^7	$2-\tau$	S_{10}^3, S_{10}^7	$-\tau$
C_{12}^1, C_{12}^{11}	$1+\sqrt{3}$	S_{12}^1, S_{12}^{11}	$-1+\sqrt{3}$
C_{12}^5, C_{12}^7	$1-\sqrt{3}$	S_{12}^5, S_{12}^7	$-1-\sqrt{3}$
C_{14}^1, C_{14}^{13}	$+1+2\cos\frac{\pi}{7}$	S_{14}^1, S_{14}^{13}	$-1+2\cos\frac{\pi}{7}$
C_{14}^3, C_{14}^{11}	$+1+2\cos\frac{3\pi}{7}$	S_{14}^3, S_{14}^{11}	$-1+2\cos\frac{3\pi}{7}$
C_{14}^5, C_{14}^9	$+1+\cos\frac{5\pi}{7}$	S_{14}^5, S_{14}^9	$-1+2\cos\frac{5\pi}{7}$
C_{16}^1, C_{16}^{15}	$1+(2+\sqrt{2})^{\frac{1}{2}}$	S_{16}^1, S_{16}^{15}	$-1+(2+\sqrt{2})^{\frac{1}{2}}$
C_{16}^3, C_{16}^{13}	$1+(2-\sqrt{2})^{\frac{1}{2}}$	S_{16}^3, S_{16}^{13}	$-1+(2-\sqrt{2})^{\frac{1}{2}}$
C_{16}^5, C_{16}^{11}	$1-(2-\sqrt{2})^{\frac{1}{2}}$	S_{16}^5, S_{16}^{11}	$-1-(2-\sqrt{2})^{\frac{1}{2}}$
C_{16}^7, C_{16}^9	$1-(2+\sqrt{2})^{\frac{1}{2}}$	S_{16}^7, S_{16}^9	$-1-(2+\sqrt{2})^{\frac{1}{2}}$

for proper rotations and

$$\chi_{\text{vib}}\left(S_n^k\right) = (n_R)(-1+2\cos\theta) \qquad (1.32\text{b})$$

for improper rotations.

The contributions of each operation to character are tabulated for convenience and they are given in Table 1.3 (J.A. Salthouse and M.J. Ware. Cambridge University Press, 1972).

Table 1.4 Revised character table

D_{2h}	E	$C_2(z)$	$C_2(y)$	$C_2(x)$	i	$\sigma(xy)$	$\sigma(xz)$	$\sigma(yz)$	μ'	α'
A_g	+1	+1	+1	+1	+1	+1	+1	+1		$\alpha'_{xx}, \alpha'_{yy}, \alpha'_{zz}$
B_{1g}	+1	+1	−1	−1	+1	+1	−1	−1		α'_{xy}
B_{2g}	+1	−1	+1	−1	+1	−1	+1	−1		α'_{xz}
B_{3g}	+1	−1	−1	+1	+1	−1	−1	+1		α'_{yz}
A_u	+1	+1	+1	+1	−1	−1	−1	−1		
B_{1u}	+1	+1	−1	−1	−1	−1	+1	+1	μ'_z	
B_{2u}	+1	−1	+1	−1	−1	+1	−1	+1	μ'_y	
B_{3u}	+1	−1	−1	+1	−1	+1	+1	−1	μ'_x	
X_{vib}	66	2	2	−2	0	24	4	0		
n_R	24	0	0	4	0	24	4	0		

We can rewrite the character table adding the character for the normal modes of vibration calculated using Equations (1.32) and Table 1.3, giving Table 1.4.

The number of normal modes a_i of each irreducible representation Γ_i is calculated as follows:

$$a_i = \frac{1}{h} \sum_R g_R \chi_i (R)\, \chi_{\text{vib}} (R) \tag{1.33}$$

where h is the order of the group and g_R is the number of operations in the Rth class. The last two factors in Equation (1.33) are the character of the irreducible representation and the character of the normal modes, respectively.

For instance, for the totally symmetric modes, the number of normal modes is

$$a_{A_g} = \frac{1}{8}(66 \times 1 + 2 \times 1 + 2 \times 1 - 2 \times 1 + 24 \times 1 + 4 \times 1) = 12,$$

and the total representation is found to be

$$\Gamma = 12a_g + 11b_{1g} + 6b_{2g} + 4b_{3g} + 5a_u + 6b_{1u} + 11b_{2u} + 11b_{3u}. \tag{1.34}$$

Lower-case letters are used for species of normal vibrational modes according to IUPAC recommendations. The results obtained for the fixed molecule with an equilibrium geometry belonging to the point group

D_{2h} allows one to know their activity in the Raman and infrared spectra. The caveat is to make sure that the Cartesian coordinates used for the molecular system correspond to the system of coordinates used in the character table provided. Most character tables given in textbooks follow Mullikan's recommendations and notation. Character tables provide the species of symmetry for dipole moment derivatives μ_i' and polarizability derivatives α_{ij}', in the last two columns of the table. Since anthracene is a centrosymmetric molecule, the mutual exclusion rule applies, and infrared-active modes are not Raman active and vice versa. Therefore, the infrared spectrum is given by

$$\Gamma_{\mathrm{IR}} = 6b_{1u} + 11b_{2u} + 11b_{3u} \quad \text{and} \quad (5a_u)$$

which are silent modes.

The Raman spectrum can contain the following active normal modes:

$$\Gamma_{\mathrm{Raman}} = 12a_g + 11b_{1g} + 6b_{2g} + 4b_{3g}.$$

This concludes our discussion on the vibrational spectrum using symmetry. The number and activity of the fundamental vibrational frequencies of each symmetry species are known. However, we have no information about the intensity with which each normal mode will be observed. The intensity of the infrared and Raman spectra can be computed *ab initio*, and this task will be shown with one more example before the conclusion of the present chapter.

1.7 SELECTION RULES

The conservation of angular momentum and parity impose restrictions on the quantum transitions of a molecule. These restrictions are collectively known as selection rules (reference 4, p. 294). In infrared spectroscopy, using the harmonic approximation, the relevant rules are the electric dipole selection rules. The description of the absorption of light by a molecule requires knowledge of the coupling of the electric dipole to an external electromagnetic field: $H' = -p \cdot E$. The probability for the absorption is therefore proportional to the square of the dipole moment matrix element along the direction $\boldsymbol{E}_j$ of light polarization. The amplitude of the transition is proportional to the scalar product $j \cdot \langle \psi_v \,|p|\, \psi_{v'} \rangle$. The selection rules for transitions between vibrational levels $\psi_{v'}$ and ψ_v

are determined by the matrix element $\langle\psi_v|p|\psi_{v'}\rangle$. Symmetry reduces the electric dipole selection rules to the requirement of equal irreducible representations of the normal mode and one of the coordinates in $\boldsymbol{p}$.

The general *selection rule* for an allowed transition between two electronic or vibrational states connected by an operator $\boldsymbol{p}$ requires that the direct product (triple product) has a totally symmetric component: $\Gamma_{\text{state}} \times \Gamma_p \times \Gamma_{\text{state}'}$ = totally symmetric (reference 6, p. 129). For an isolated molecule or gas-phase spectrum, the triple product is directly given in the character table as we described in the example in Section 1.6. However, when there is a molecular orientation as in solids, adsorbates and films or low-temperature experiments, the scalar product $j \cdot \langle\psi_v|p|\psi_{v'}\rangle$ becomes the most important tool in the spectral interpretation of the observed intensities. Standard point groups can be used for adsorbed molecules and surface complexes, for which we can ignore all but one equilibrium configuration. The reduced representation of the μ' and α'_{ij} operators connecting spectroscopic infrared and Raman transitions are listed in the character table. For example, the electric dipole moment operator μ' transforms as x, y, z where as the electric polarizability α'_{ij} (second rank tensor) operator transforms as $x^2, y^2, z^2, xy, xz, yz$. For crystals [12], adsorbates or thin solid films, there may exist spatial anisotropy introduced by molecular alignment. Therefore, the observed intensity of the allowed infrared and Raman modes can be modulated by a well-defined spatial orientation (polarization) of the incident electric field. This means that allowed infrared modes of a given symmetry species will be seen with an absorption intensity proportional to the square of the scalar product, $\mathrm{E} \cdot \mu'$, i.e. the square of the cosine of the angle between the polarization of the vector E and the directional properties of the dynamic dipole μ' $(\partial\mu/\partial Q)$. The corollary is that symmetry species of single crystals and adsorbed molecules of known orientation may be distinguished by the use of polarized radiation [12]. For the interaction Hamiltonian, the most notable practical applications are the following:

1. To describe infrared experiments on molecules adsorbed on reflecting metal surfaces one follows the realization that the light at the reflecting surface is highly polarized and, at the appropriate angle of incidence, the p-polarized component of the electromagnetic wave is three orders of magnitude larger than the parallel component. The latter is the basis for the polarization selection rules of specular reflection–absorption infrared spectroscopy (RAIRS). As a result, RAIRS is the most extensively used technique to determine the orientation of nanometric organic films [17,18].

2. To describe Raman experiments on single crystals using plane polarized incident light, one follows the convention of Damen, Porto and Tell (Porto's notation) [19].
3. To describe Raman experiments on molecules adsorbed on metal surfaces using plane polarized radiation, one follows the convention of surface selection rules (propensity rules) [20,21].

Practical applications of the selection rules will be given in the corresponding chapters where we discuss reflection–absorption infrared spectroscopy (RAIRS), surface-enhanced Raman scattering (SERS) and surface-enhanced infrared absorption (SEIRA). However, we finish this chapter with an example of Raman and infrared intensity calculations with special attention to the allowed intensities for an oriented molecule.

1.8 THE EXAMPLE OF *AB INITIO* COMPUTATION OF THE RAMAN AND INFRARED SPECTRA

To illustrate the power of computational chemistry in vibrational spectroscopy, we take as a study case 3,4,9,10-perylenetetracarboxylic acid dianhydride (PTCDA) [22], a planar D_{2h} molecule ($C_{24}H_8O_6$, MW 392.347). There are 108 normal modes and the total irreducible representation calculated using the method described in Section 1.6 is

$$\Gamma = 19a_{1g} + 7b_{1g} + 10b_{2g} + 18b_{3g} + 8a_u + 18b_{1u}(z) + 18b_{2u}(y) + 10b_{3u}(x).$$

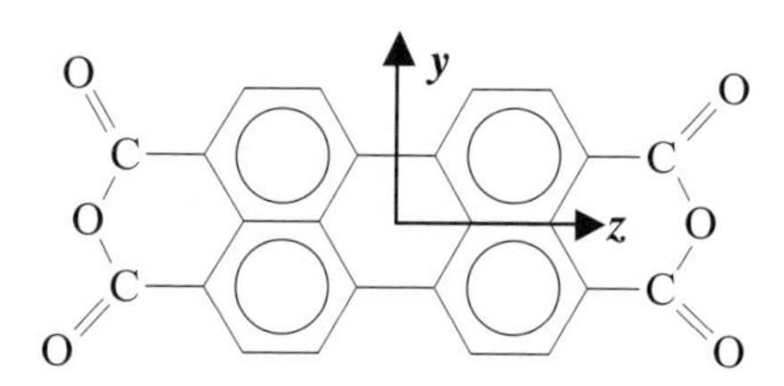

The procedures employed use the character table and follow the convention for the selection of the molecular axes for the D_{2h} point group. The conventions for other groups are listed at the end of this section. Therefore, the x-axis is perpendicular to the molecular plane and the z-axis passes through the greatest number of atoms. The latter convention should be strictly followed to maintain correspondence between the species of the total irreducible representation Γ and those given in the character table. This is of paramount importance when studying molecular orientation on surfaces or solid-state materials. One of the most commonly encountered applications is in RAIRS or reflection Raman scattering from thin films on reflecting surfaces. Correspondingly, the same convention should be followed when interpreting SEIRA or SERS spectra. We will use here the infrared spectra of PTCDA to demonstrate the analytical applications of the quantum chemical computations. The PTCDA molecule was computed using Gaussian 98 [23] at the B3LYP/6–31G(d) level of theory. The initial geometry was minimized within a PM3 calculation, and the corresponding checkpoint file was used as a starting point in the DFT [B3LYP/6–31G(d)] computation. The calculated infrared spectra are compared with the infrared spectrum of the solid PTCDA dispersed in a KBr matrix. The calculated spectra should correspond closely to the infrared spectrum of PTCDA in the gas phase. There can be striking differences between the solid-state and the gas-phase spectra. The quantum chemical calculated frequencies are scaled, and the scaled wavenumbers agree better with the observed spectra. The top spectrum is presented without the scaling factor and the bottom computed frequencies have been scaled using the same factor for the entire spectrum, equal to 0.9614. The rationale and introduction of the numerical scaling factors are discussed in specialized reports [24] and the benefits of scaling can be seen in Figure 1.3.

Notably, the infrared spectrum as shown is composed of $8a_u + 18b_{1u}(z) + 18b_{2u}(y) + 10b_{3u}(x)$. However, the eight a_u are silent and there are 46 active fundamentals to be observed. The $10b_{3u}$ modes are observed in the spectral range below 1000 cm^{-1}; they are the out-of-plane vibrations that will include twisting or torsion vibrational modes (τ) and wagging modes (ω). The wagging corresponds to changes in dihedral angles, angles between molecular planes. The $18b_{1u}(z) + 18b_{2u}(y)$ infrared-active fundamentals contain bond stretching (ν) modes and deformation or angle bending (δ) modes. In summary, the infrared spectrum shown in Figure 1.3 is the sum of three spectra, $18b_{1u}(z) + 18b_{2u}(y) + 10b_{3u}(x)$, and these spectra could be observed independently for fixed molecular orientation and using the appropriate polarized light. The latter will be demonstrated as an application of polarized infrared spectroscopy in the next chapter. The computed spectra separated by symmetry species are given in Figure 1.4. The same approach applies to computed Raman spectra. Each of the species of symmetry is plotted independently, and therefore the relative intensities are seen with respect to the strongest infrared active mode within the group. For plotting purposes, each calculated wavenumber has been given a full width at half-maximum (FWHM) of 5 cm^{-1}.

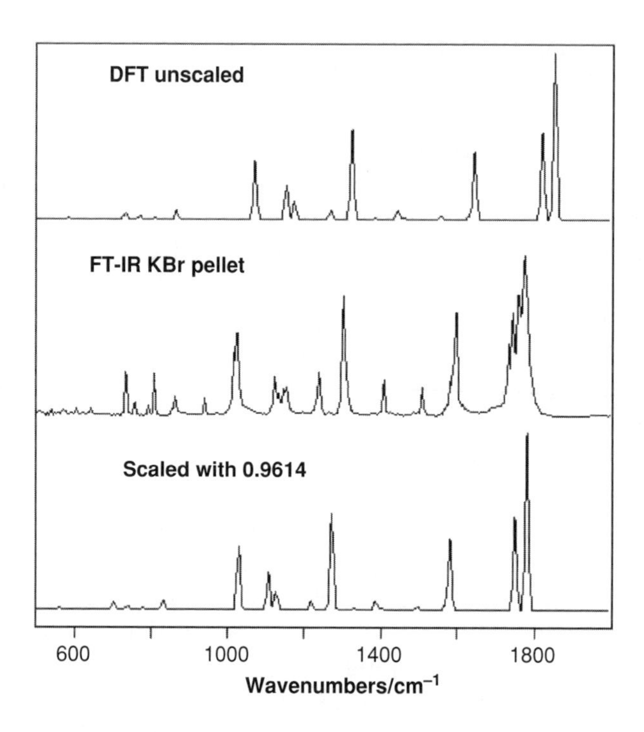

Figure 1.3 DFT [B3LYP/6–31G(d)]-calculated (unscaled and scaled) infrared spectra of PTCDA compared with the infrared spectrum of the solid material dispersed in a KBr matrix

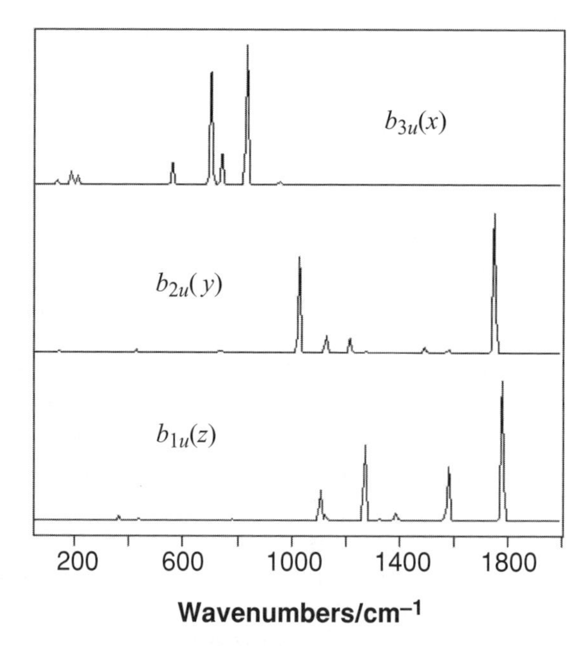

Figure 1.4 DFT-calculated spectra of infrared-active symmetry species in PTCDA

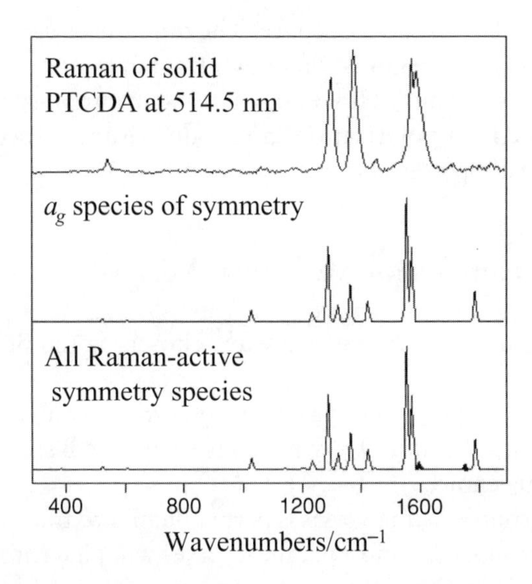

Figure 1.5 Raman spectrum of solid PTCDA recorded with the 514.5 nm laser excitation. DFT-calculated a_{1g} Raman spectrum, and computed spectrum with the sum of all allowed Raman fundamental vibrational modes

It can be speculated that for a perfectly oriented PTCD molecule, infrared light polarized along x will produce the top spectrum in Figure 1.4. Similarly, infrared light polarized along y would give the middle infrared spectrum and finally the z-polarized infrared light will deliver the spectrum at the bottom. This explains the rapid development of polarization spectroscopy in solid state vibrational spectroscopy and surface vibrational techniques [12,25].

The Raman spectrum was calculated and the Raman-active fundamentals that can be observed are $19a_{1g} + 7b_{1g} + 10b_{2g} + 18b_{3g}$, or 54 bands can be recorded. The spectrum, recorded with 514.5 nm laser radiation, of a solid sample of PTCDA is shown in Figure 1.5 (top spectrum). The calculated spectrum, including all Raman-active species of symmetry, i.e. 54 fundamentals, is plotted at the bottom, for comparison. The agreement here is remarkable, despite the fact the excitation is near resonance (resonance Raman scattering) and the intensities may deviate considerably from those of the normal Raman scattering. The middle spectrum corresponds to the a_g species containing only 19 fundamental vibrational modes. It can be seen that the totally symmetric modes a_g are responsible for the intensity pattern observed in the Raman spectrum of PTCDA, and the other species have minor relative intensity in the Raman spectrum averaged over all directions of space. The differences can be identified in the 1600–1800 cm^{-1} region, where two Raman bands of weak intensity are seen in the spectrum. However, as in the infrared, spatial molecular orientation and light polarization can produce Raman spectra with quite different intensity patterns [19].

1.8.1 Conventions for Molecular Axes

In the axial groups the z-axis is always chosen as the principal axis of symmetry.

The selection of the x-and y-axes in some groups remains arbitrary. In point groups C_{nv}, D_n and D_{nh} (where n is even), the B species are directly affected by this choice.

For point group C_{2v}, the x-axis is perpendicular to the molecular plane.

For point groups D_{4h} and D_{6h}, the C_2 axes will pass through the greatest number of atoms (or intersect the largest number of bonds).

For point groups C_{nv} with even n, the σ_v plane will pass through the greatest number of atoms (or intersect the largest number of bonds). To define x and y in these groups they may be chosen so that the x-axis lies in one of the σ_v planes.

1.9 VIBRATIONAL INTENSITIES

The probability of the dipole transition is proportional to the square of the magnitude of the transition dipole moment (or 'dynamic dipole') [3,7,12]. Working within the ground electronic state, i.e. during the transition there is no change in the molecular electronic states, the vibrational transition moments are defined by [3,7] $|\mu'|^2 = |(\mu)_{v'v''}|^2 = |\langle v'|\mu|v''\rangle|^2$ for transitions in the infrared spectra and by $|\alpha'|^2 = |(\alpha)_{v'v''}|^2 = |\langle v'|\alpha|v''\rangle|^2$ for transitions observed in the Raman spectra. Dipole moment and polarizability undergo infinitesimal changes during molecular vibrations. Near the equilibrium geometry, both molecular properties can be expanded as a Taylor series in the normal coordinates. Therefore, we write again here the expansion for the dipole moment and the polarizability:

$$p = \mu \text{ or } \alpha = p_0 + \sum_{i}^{n} \left(\frac{\partial p}{\partial Q_i} \right) Q_i + \frac{1}{2} \sum_{i,j}^{n} \left(\frac{\partial^2 p}{\partial Q_i \partial Q_j} \right) Q_i Q_j + \ldots \tag{1.35}$$

Where Q_i is the ith normal coordinate belonging to the set of Q that diagonalizes the potential energy function and is associated with the harmonic frequency ω_i. The expressions (1.35) are used to calculate the transition dipole moments in $|\mu'| = |\langle v'|\mu|v''\rangle|$ and $|\alpha'| = |\langle v'|\alpha|v''\rangle|$. Since the functions of the harmonic oscillator are orthonormal, in the harmonic approximation the first term in Equation (1.35) gives zero for μ, and gives rise to the elastic Rayleigh scattering for α. In the harmonic approximation, the third term in both series is neglected and, thereby, the infrared and Raman spectra are completely determined by the first derivatives of the dipole moment and first derivatives of the polarizability. The calculation of the fundamental non-zero transition dipole moment is simply equal to

$$|\mu'| = |\langle v'| \left(\frac{\partial \mu}{\partial Q_i}\right) Q_i \; |v''\rangle| = \left(\frac{\partial \mu}{\partial Q_i}\right)\left(\frac{h}{4\pi\omega_i}\right)^{\frac{1}{2}}. \tag{1.36}$$

When all the matrix elements are taken into account in the transition dipole, the final result is

$$|\mu'| = \left(\frac{\partial \mu}{\partial Q_i}\right)\left(\frac{h}{8\pi^2 \nu_i}\right)^{\frac{1}{2}} (v_i + 1)^{\frac{1}{2}} \tag{1.37}$$

and its squared magnitude is given by

$$|\mu'|^2 = \left(\frac{\partial \mu}{\partial Q_i}\right)^2 \left(\frac{h}{8\pi^2 \nu_i}\right) (v_i + 1)\,. \tag{1.38}$$

Experimentally, the definition of the absorptivity is usually taken from the equation for the exponential attenuation for irradiance (Lambert's law): $I = I_0 e^{-\kappa z}$, where I_0 is the incident irradiance and I is the irradiance at depth z of the absorber. The frequency-dependent κ(kappa) is the extinction coefficient, sometimes called the absorption coefficient, and is a characteristic property of the material through which the light is passing. Let us consider the irradiance or radiant flux per unit of area of surface I (in W/m^{-2}) of a monochromatic beam of frequency ν traveling along the $+z$ direction with irradiance which is equal to the energy density multiply by its speed: $I_0(z) = c \times \rho(\nu, z)$, c being the speed of light. For the transition in a two-state system, with a density of oscillators in the lower state equal to N_1, the number of photons absorbed in 1 within a distance dz and unit area is $B_{1,2}(\nu) N_1 \rho(\nu,z) dz$. Each absorption takes an $h\nu$ amount of energy from the beam, and thereby the change in the irradiance is [7]:

$dI = -h\nu_{1,2}B_{1,2}(\nu)(N_1 - N_2)\rho(\nu,z)dz$. Using the expression for the initial irradiance I_0, the change is

$$dI = -I_0 \frac{h\nu_{1,2}B_{1,2}(\nu)N_1}{c} dz.$$

Assuming that the attenuation is entirely due to absorption, the absorption coefficient is

$$\kappa(\nu) = \frac{h\nu_{1,2}B_{1,2}(N_1 - N_2)}{c}.$$

Since the Einstein coefficient for the absorption of a vibrating molecule (without rotations) is given in terms of the transition dipole moment [3,7]:

$$B_{v'v''} = \frac{8\pi^3}{3h^2}[|\langle v'|\mu_x|v''\rangle|^2 + |\langle v'|\mu_y|v''\rangle|^2 + |\langle v'|\mu_z|v''\rangle|^2]$$

or in the SI system of units

$$B_{v'v''} = \frac{1}{6\varepsilon_0\hbar^2}|\langle v'|\mu|v''\rangle|^2.$$

With the help of Equation (1.38), an expression is found for the absorption coefficient using $B_{v'v''}$:

$$\kappa(\nu) = \frac{8\pi^3}{3ch}\nu_{1,2}(N_1 - N_2)|\langle v|\mu|v\rangle|^2. \tag{1.39}$$

Since the experimental infrared intensity is not a line, but a band with a well-defined FWHM, the corresponding integrated absorption coefficient is given by $A = \int_{\text{Band}} \kappa(\nu)d\nu$. In practice, the quantity A is used for the determination of $(\partial\mu/\partial Q_i)^2$. When the absorbance is proportional to concentration, the Beer extension of the Lambert law (Beer–Lambert law) can be formulated. The absolute infrared intensity of an absorption band is given by the integration over the band, at some standard pathlength

(l) and molecular concentration (c) of the sample [26]:

$$A = \frac{1}{cl}\int_{\text{Band}} \ln\left(\frac{I_0}{I}\right) d\nu. \tag{1.40}$$

Absorbance, is a term recommended for use with this measurement, in preference to absorbancy, optical density or simply extinction. Equating Equation (1.39) to the integrated absorption, at temperatures where $N_1 - N_2 = N_1$, the final results is

$$A = \frac{\pi N}{3c} \times \left[\left(\frac{\partial \mu_x}{\partial Q_i}\right)^2 + \left(\frac{\partial \mu_y}{\partial Q_i}\right)^2 + \left(\frac{\partial \mu_z}{\partial Q_i}\right)^2\right] \tag{1.41}$$

It is therefore necessary to specify the units of concentration and length. The most commonly used units in chemistry are mol L^{-1} and cm. The product concentration (moles cm^{-3}) and pathlength (cm), shown as *cl* in Equation (1.40), has the units of mol cm^{-2}. The integral has dimensions of cm^{-1}, when the variable for the integration is wavenumber. The units for A are then cm^2 mol^{-1}/cm^{-1} or cm mol^{-1}. However, these units produce large numerical values, on the order of 10^6 for fundamentals. By changing cm to km (multiplying by 10^{-5}), the intensities can be expressed by numbers in the range from 0 to 10^2. At present, the absolute integrated intensities A are commonly reported in km mol^{-1} and these units are also used in quantum chemical computations of infrared intensities. Two alternative units to consider are the 'dark' and the 'intensity unit', where 1 dark = 10^3 cm mol^{-1} and the 'intensity unit' = 10^7 cm mol^{-1} [26]. A thorough discussion of the experimental units and conversion factor can be found in a review by Pugh and Rao [27], and for liquid-state band intensities in Ratajczak and Orville-Thomas [28]. Conversion factors are given in Table 1.5.

Table 1.5 Conversion factors to km mol^{-1} [11]

Intensity unit for A	Concentration units	Conversion factor
cm mol^{-1}	mol cm^{-3}	10^{-5}
cm/$mmol^{-1}$	mol L^{-1}	10^{-2}
cm^{-2} atm (273/K)	atm	0.224
cm^{-2} atm, (298/K)	atm	0.245
s^{-1}/cm^{-1}/atm^{-1}	atm	7.477×10^{-12}

Working within the harmonic approximation, the computation of the magnitude of the dipole moment derivatives can be directly evaluated from the integrated intensities:

$$\left(\frac{\partial \mu}{\partial Q_i}\right)^2 = \frac{3c}{\pi N} \times A \tag{1.42}$$

where the constant depends on the units selected for the calculation. There is, however, an alternative to the integrated absorption coefficient, which has been recommended to be used for work on infrared intensities, namely the integral

$$\Gamma = \frac{1}{cl} \int_{\text{Band}} \ln\left(\frac{I_0}{I}\right) \text{dln}\nu. \tag{1.43}$$

The relationship between the two integrals is $A = \Gamma \times \nu_i$, where ν_I is the band center. The corresponding dipole moment derivative is related to Γ:

$$\left(\frac{\partial \mu}{\partial Q_i}\right)^2 = \frac{3c\nu_i}{\pi N_0} \times \Gamma. \tag{1.44}$$

In the presence of degeneracy, the right-hand side should be multiplied by the degeneracy factor. An important practical difference is that the units for Γ are cm^{2}mol^{-1}, where as the units of A are cm mol^{-1}. The attractive simplicity of A is in the fact that it is directly proportional to the dipole moment derivative:

$$\left(\frac{\partial \mu}{\partial Q_i}\right)^2 = \frac{3c}{\pi N_0} \times A. \tag{1.45}$$

If A_n is given in km/mol^{-1}, then the absolute value of the dipole moment derivative is given by

$$\left(\frac{\partial \mu}{\partial Q_n}\right)_0 = 0.0320 \times \sqrt{A_n}. \tag{1.46}$$

The HCl band at 2886 cm^{-1} has an integrated absorption $A = 33.2$ km/mol^{-1}. For the commonly used units of electric dipole moment, debye, the conversion factor is 1D $= 3.3356 \times 10^{-30}$ Cm. Correspondingly, the units of the dipole moment derivatives are 1 D Å^{-1} $= 3.3356 \times 10^{-20}$ N$^{\frac{1}{2}}$ C g$^{-\frac{1}{2}}$ $= 0.2083$ e(amu)$^{-\frac{1}{2}}$ and 1 cm$^{\frac{3}{2}}$ s^{-1} $= 2.684 \times 10^{-3}$ e(amu)$^{-\frac{1}{2}}$.

Table 1.6 Infrared optical cross-sections for fundamental vibration bands of methane and ethane (see Gussoni, in reference 11)

Molecule	Wavenumbers/cm^{-1}	$\Gamma/cm^2\ mol^{-1}$	Cross-section/cm^2
CH_4	3019	2310	3.83×10^{-21}
	1306	2554	4.24×10^{-21}
CH_3CH_3	2974	4113	6.83×10^{-21}
	2915	1640	2.72×10^{-21}
	1460	910	1.51×10^{-21}

Another measure of intensity is the 'integrated optical cross-section' which can be obtained from the A values in km mol^{-1} by dividing by the factor 6.022×10^{20}, or the most commonly used 'optical cross-section', obtained by dividing the Γ values (in cm^2mol^{-1}) by 6.022×10^{23}.

Infrared cross-sections for the fundamental bands of methane and ethane are given in Table 1.6.

1.9.1 Raman Intensities

$$p = \alpha E$$

where E = electric field ($N\,C^{-1}$ or $V\,m^{-1}$) and p = induced electric dipole moment of a molecule, which is a vector pointing from the negative to the positive charge ($C\ m^{-1}$) the unit debye (D) is also used: $1\,D = 3.336 \times 10^{-30}$ C m. Using C V = J, α, the polarizability has the units $C^2\ m^2\ J^{-1}$ or m^3. For instance, the average polarizability of CCl_4 is $10.5 \times 10^{-10}\ m^3$.

Absolute Raman intensities are reported in terms of the polarizability derivative α'. For instance, the derivative of the polarizability with respect to the normal coordinate for the 2914 cm^{-1} band of methane has been reported to be [29], $45\alpha^2 = 190 \cdot N_a \cdot 10^{-32}\, cm^4 \cdot g^{-1}$ (or 190 $Å^4\ amu^{-1}$). The Raman intensities in quantum chemistry computations are commonly reported in $Å^4\ amu^{-1}$, angstroms (10^{-10} meter or also known as tenthmeter) and unified atomic mass units, $1.660531\ 10^{-24}$ g.

1.10 DEFINITION OF CROSS-SECTION

Experimentally, the spontaneous inelastic Raman scattering (RS), the total Stokes scattered light, averaged over all random molecular orientations, I_{RS} (photons s^{-1}), is proportional to the incoming flux of photons, I_0 (photons s^{-1} cm^{-2}): $I_{RS} = \sigma_{RS} I_0$.

The proportionality constant, the Raman cross-section σ_{RS}, has the dimensions of cm^2 and is a function of the frequency of excitation. The Raman cross-section is proportional to the square of the polarizability derivative for the $m \rightarrow n$ vibrational transition, $\alpha' = (\partial\alpha/\partial Q)_0$, and the fourth power of the scattering frequency ω_S: $\sigma_{RS} = C\omega_S{}^4|\alpha'_{mn}|^2$. C contains numerical constants.

The efficiencies of the absorption and scattering processes are determined by the function 'cross-section', which is the meeting point of experiments with theory. There are three common quantities used by spectroscopists. The spectral differential cross-section is the rate of removal of energy from the light beam into a solid angle dΩ and frequency interval dω: $d^2\sigma/d\Omega d\omega$. Integration restricted to include a single intensity peak gives the differential cross-section $d\sigma/d\Omega$.

Integration of the differential cross-section over all directions in space gives the cross-section σ. The units of σ are m^2, and the cross-section can be interpreted as the target area presented by a molecule (particle) for scattering or absorption. The definition is easily extended to emission processes. Typical values of the cross-section for the spectroscopic processes of interest are illustrated in the Figure 1.6, and specific values for given molecular vibrations can be found in modern books [30] with reference to the original work. For instance, the cross-section σ for absorption in the infrared is ca 10^{-20} cm^2. Therefore, in Figure 1.6, the value assigned to the y-axis for the infrared cross-section is $-\log(10^{-20}) = 20$. The absolute Raman cross-section for the 666 cm^{-1} mode of

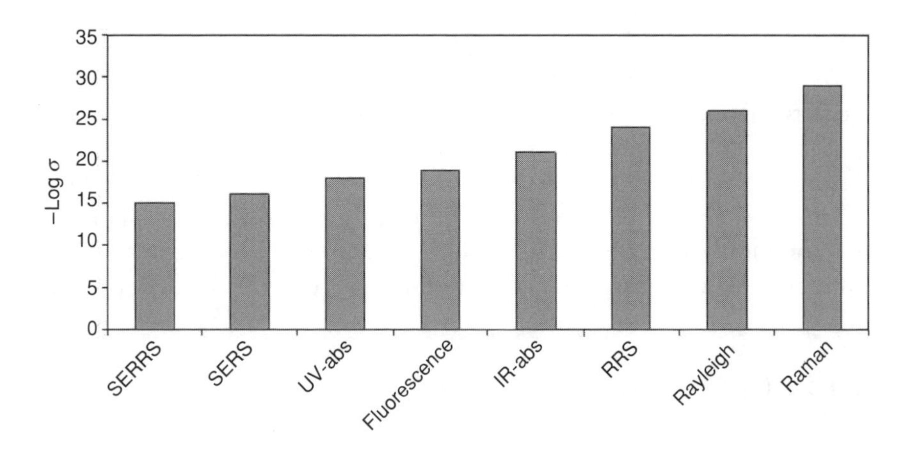

Figure 1.6 A plot of $-\log\sigma$ (cross-section in cm^2 per molecule) for the most common optical processes in linear spectroscopy

Table 1.7 Excitation wavelength λ (nm) and σ_R ($\times 10^{28}$ cm^2 or $\times 10^{12}$ Å^2).

λ	$\sigma_R{}^a$
532.0	0.660 ± 0.1
435.7	1.660 ± 0.5
368.9	3.760 ± 0.3
355.0	4.360 ± 0.4
319.9	7.560 ± 0.3
282.4	13.06 ± 4.0

[a] After Foster *et al.* [31]. Error represents one standard deviation from the mean.

Table 1.8 Approximate order of magnitude for cross-sections σ (per molecule) for various possible processes in spectroscopy

Process	Cross-section of	σ/cm^2
Absorption	Ultraviolet	10^{-18}
Absorption	Infrared	10^{-21}
Emission	Fluorescence	10^{-19}
Scattering	Rayleigh scattering	10^{-26}
Scattering	Raman scattering	10^{-29}
Scattering	Resonance Raman	10^{-24}
Scattering	SERRS	10^{-15}
Scattering	SERS	10^{-16}

chloroform [31] has been determined using several-laser lines. The results are given in Table 1.7.

The approximate cross-sections for the most common spectroscopies are listed in Table 1.8.

The estimated SERS cross-section is taken from the review by Kneipp *et al.* [32]. The SERRS cross-section is the one reported by Nie and Emory [33] for Rhodamine 6G (R6G) excited in resonance at 514.5 nm. Later, Michaels *et al.* [34] reported an average SERRS cross-section for R6G of 2×10^{-14} cm^2 at 514.5 nm. The SERS/SERRS cross-sections correspond to experimental results with the best observed enhancement factors.

1.11 THE UNITS OF ENERGY AND FORCE CONSTANTS

Force constant: $k_e = 4\pi^2 \mu c^2 \omega^2$, where ω is the molecular vibration in wavenumbers (cm^{-1}) and $4\pi^2 \mu c^2$ is the classical force constant factor: $4\pi^2\mu c^2 = 5.98180 \times 10^{-9}$ N m^{-1} = 5.98180×10^{-6} N cm^{-1} = 5.98180×10^{-6} myn Å^{-1}.

Energy (hartree) (used in Gaussian 98) $= (2\pi)^4 m_e e^4/h^4 = 27.2115$eV $= 4.3598138 \times 10^{-18}$ J.

REFERENCES

[1] G. Herzberg, *Molecular Spectra and Molecular Structure. I. Infrared and Raman Spectra of Polyatomic Molecules*, Van Nostrand, Princeton, NJ.

[2] G. Herzberg, *Molecular Spectra and Molecular Structure. II. Spectra of Diatomic Molecules*. Van Nostrand, Princeton, NJ, 1950.

[3] E.B. Wilson, Jr J.C. Decius and P.C. Cross, *Molecular Vibrations; the Theory of Infrared and Raman Vibrational Spectra*, McGraw-Hill, New York, 1955.

[4] A.S. Davydov, *Quantum Mechanics*, Pergamon Press, Oxford 1961.

[5] R.P. Feynman, R.B. Leighton and M. Sands, *The Feynman Lectures on Physics*, Addison-Wesley, Reading, MA, 1965.

[6] G. Herzberg, *Molecular Spectra and Molecular Structure. III. Electronic Spectra and Electronic Structure of Polyatomic Molecules*. Van Nostrand, Princeton, NJ, 1966.

[7] M.B. Bolkenshtein, L.A. Gribov, M.A. Eliashevich and B.I. Stepanov, *Molecular Vibrations*, Nauka, Moscow, 1972.

[8] J.C. Tully Chemical dynamics at metal surfaces. *Annu. Rev. Phys. Chem* 2000, **51**, 153–178.

[9] M. Diem, *Modern Vibrational Spectroscopy*, John Wiley & Sons, Inc., New York, 1993.

[10] L.A. Woodward, *Introduction to the Theory of Molecular Vibrations and Vibrational Spectroscopy*, Oxford University Press, Oxford, 1972.

[11] W.B. Person and G. Zerbi, *Vibrational Intensities in Infrared and Raman Spectroscopy*, Elsevier, New York, 1982.

[12] J.C. Decius and R.M. Hexter, *Molecular Vibrations in Crystals*, McGraw-Hill, New York, 1977.

[13] D.A. Long, *The Raman Effect*, John Wiley & Sons, Ltd, Chichester, 2001.

[14] K. Nakamoto, *Infrared and Raman Spectra of Inorganic and Coordination Compounds. Part A: Theory and Applications in Inorganic Chemistry*, John Wiley & Sons, Inc., New York, 1997.

[15] J.R. Ferraro and J.S. Ziomek, *Introductory Group Theory and Its Applications to Molecular Structure*, Plenum, Press, New York, 1969.

[16] F.A. Cotton, *Chemical Applications of Group Theory*, John Wiley & Sons, Inc., New York, 1963.

[17] B.E. Hayden, in J.T. Yates Jr and T.E. Madey (eds), *Vibrational Spectroscopy of Molecules on Surfaces, Methods of Surface Characterization*, Plenum Press, New York, 1987, pp. 267–340.

[18] M.K. Debe, Optical probes of organic thin films, *Prog. Surf. Sci.* 1987, **24**, 1–281.

[19] T.C. Damen, S.P.S. Porto and B. Tell, *Phys. Rev.* 1966, **142**, 570.

[20] M. Moskovits, Surface selection rules. *J. Chem. Phys.* 1982, **77**, 4408–4416.

[21] A. Campion, in J.T. Yates Jr and T.E. Madey, (eds), *Vibrational Spectroscopy of Molecules on Surfaces, Methods of Surface Characterization*, Plenum Press, New York, 1987, pp. 345–412.

[22] K. Akers, R. Aroca, A.M. Hor and R.O. Loutfy, Molecular organization in perylene tetracarboxylic dianhydride films, *J. Phys. Chem.* 1987, **91**, 2954–2959.

[23] M.J. Frisch, G.W. Trucks, H.B. Schlegel, G.E. Scuseria, M.A. Robb, J.R. Cheeseman, V.G. Zakrzewski, J.A. Montgomery Jr, R.E. Stratmann, J.C. Burant, S. Dapprich, J.M. Millam, A.D. Daniels, K.N. Kudin, M.C. Strain, O. Farkas, J. Tomasi, V. Barone, M. Cossi, R. Cammi, B. Mennucci, C. Pomelli, C. Adamo, S. Clifford, J. Ochterski, G.A. Petersson, P.Y. Ayala, Q. Cui, K. Morokuma, D.K. Malick, A.D. Rabuck, K. Raghavachari, J.B. Foresman, J. Cioslowski, J.V. Ortiz, B.B. Stefanov, G. Liu, A. Liashenko, P. Piskorz, I. Komaromi, R. Gomperts, R.L. Martin, D.J. Fox, T. Keith, M.A. Al-Laham, C.Y. Peng, A. Nanayakkara, C. Gonzalez, M. Challacombe, P.M.W. Gill, B. Johnson, W. Chen, M.W. Wong, J.L. Andres, C. Gonzalez, M. Head-Gordon, E.S. Replogle and J.A. Pople, *Gaussian 98, Revision A.3*, Gaussian, Pittsburgh, PA, 1998.

[24] A.P. Scott and L. Radom, Harmonic vibrational frequencies: an evaluation of Hartree–Fock, Møller–Plesset, quadratic configuration interaction, density functional theory, and semiempirical scale factors, *J. Phys. Chem.* 1996, **100**, 16502–16513.

[25] P.M.A. Sherwood, *Vibrational Spectroscopy of Solids*, Cambridge University Press, Cambridge, 1972.

[26] M. Mills, Infrared intensities, *Annu. Rep. Chem. Soc. London* 1958, **55**, 55–67.

[27] L.A. Pugh and N.K. Rao, in K.N. Rao (ed), *Molecular Spectroscopy: Modern Research*, Academic Press, New York, 1976, pp. 165–227.

[28] H. Ratajczak and W.J. Orville-Thomas, Infrared dispersion studies, *Trans. Faraday Soc.* 1965, **61**, 2603–2611.

[29] H.W. Schrotter and H.J. Bernstein, *J. Mol. Spectrosc.* 1961, **7**, 464.

[30] C.L. Stevenson and T. Vo-Dinh, in J.J. Laserna (ed), *Modern Techniques in Raman Spectroscopy*, John Wiley & Sons Ltd, Chichester, 1996, p. 22.

[31] C.E. Foster, B.P. Barham and P.J. Reida, Resonance Raman intensity analysis of chlorine dioxide dissolved in chloroform: The role of nonpolar solvation, *J. Chem. Phys.* 2001, **114**, 8492–8504.

[32] K. Kneipp, H. Kneipp, I. Itzkan, R.R. Dasari and M.S. Feld, Ultrasensitive chemical analysis by Raman spectroscopy, *Chem. Rev.* 1999, **99**, 2957–2975.

[33] S. Nie and S.R. Emory, Probing single molecules and single nanoparticles by surface-enhanced Raman scattering, Science 1997, **275**, 1102–1106.

[34] A.M. Michaels, M. Nirmal and L.E. Brus, Surface enhanced Raman spectroscopy of individual Rhodamine 6G molecules on large Ag nanocrystals, *J. Am. Chem. Soc.* 1999, **121**, 9932–9939.

2

The Interaction of Light With Nanoscopic Metal Particles and Molecules on Smooth Reflecting Surfaces

It is shown that the particles seen in a gold ruby glass are particles of gold which, when their diameters are less than 0.1 μ, are accurately spherical. I have endeavoured to show that the presence of many of these minute spheres to a wave-length of light in the glass will account for all the optical properties of regular gold ruby glass, and that the irregularities in colour and in polarization effects sometimes exhibited by gold glass are due to excessive distance between consecutive gold particles or to excessive size of such particles, the latter, however, involving the former.

J. C. Maxwell-Garnett (1)

The seminal work on the optical properties of nanoparticles was published a century ago by Maxwell-Garnett [1] and Mie [2]; however, the race to fabricate nanoparticles with control over the shape and size is today in full swing and there are now several books dedicated to nanoparticle fabrication, properties and applications [3–6]. Isolated particles between 1 and 100 nm in size are generally accepted under the general definition of nanoparticles. The latter general definition will include molecules and supramolecular structures (buckyballs, DNA, proteins and dendrimers); however, for the discussion of SERS/SEIRA, we

Surface-Enhanced Vibrational Spectroscopy R. Aroca

are particularly interested in metal nanoparticles large enough to support surface plasmon resonances. A thorough discussion of the cluster-size effect on the optical properties, in particular plasmon resonances, is given in Chapter 2 of Kreibig and Vollmer's book [4], where they introduced a practical classification of the clusters according to the number of atoms N. Very small clusters for $N \leq 20$, small clusters for $20 \leq N \leq 500$ and large clusters for $N \leq 10^7$. Recently, it has been shown that gold cluster of $N = 20$ can be very stable, showing chemical stability thanks to a large energy gap between the highest occupied molecular orbital (HOMO) and the lowest unoccupied molecular orbital (LUMO) [7]. Clearly, very small clusters do not have the density of states required to support plasmons; correspondingly, Kreibig and Vollmer recommend that the term plasmon should not be used for collective electronic excitations of these clusters. The transition from the molecular behavior, with discrete energy levels, to the plasmon-supporting particle, with a conduction band, is a function of the cluster size for free-electron metals, and is a very active field of research in solid-state physics [8,9]. In the present chapter we only consider nanoparticles larger than 5 nm in size, for which classical Mie theory provides correct resonance positions.

The most common morphological (size and shape) characterization techniques are high-resolution transmission electron microscopy (HRTEM) and atomic force microscopy (AFM). Examples of some HRTEM of nanoparticles fabricated in our laboratory are shown in Figure 2.1.

The interaction of light with nanoparticles spans the entire spectral region (ultraviolet to mid-infrared) relevant to infrared and Raman spectroscopy, providing the theoretical background (the optics) for the explanation and development of SERS and SEIRA. The near-and the far-field in excited nanoparticles are key elements in the background of the plasmon-assisted enhanced-optical phenomena.

> Optical near fields exist close to any illuminated object. They account for interesting effects such as enhanced pinhole transmission or enhanced Raman scattering enabling single-molecule spectroscopy. Also they enable high-resolution (below 10 nm) optical microscopy. The plasmon-enhanced near-field coupling between metallic nanostructures opens new ways of designing optical properties.
>
> **R. Hillebrand, T. Taubner and F. Kellman [10]**

The propagation and interaction of light with media is the realm of *optics*. Since the principles of optics were formulated before 1900 (before

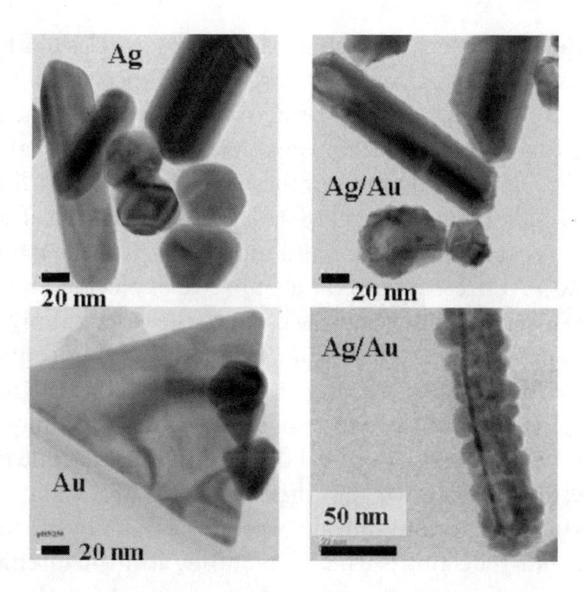

Figure 2.1 High-resolution electron microscopy (HTEM) of silver, gold and bimetallic silver–gold nanoparticles prepared from colloidal solutions

quantum mechanics), there are two fully developed approaches to optics: classical optics and quantum optics [11]. The optics needed here is restricted to linear optical techniques and the linear optical properties of metal particles and reflecting surfaces. Maxwell's phenomenological theory [12–14] is used to explain light propagation and dispersion in various media, in terms of certain optical parameters or materials 'constants': the electric permittivity (dielectric constant) ϵ, the magnetic permeability μ and the electric conductivity σ. The dependence of these parameters on the spectral frequency (dispersion theory) requires a discussion that correlates them to the atomistic structure of matter. All macroscopic phenomena of light propagation, reflection, refraction, absorption and polarization, are derivable from Maxwell's equations. In particular, the absorption and scattering of light by nanoscopic metal particles is central to the SERS effect. At the same time, reflection, absorption and scattering optical techniques are central tools for the characterization of nanomaterials, organized structures and chemically assembled nanomaterials [15]. The scattering and absorption of light by particles is largely determined by the response function of the material, the dielectric permittivity tensor, which describes the optical properties. The dependence of the absorption and scattering of metal particles on their optical properties, size and shape is required background for the discussion of SERS and SEIRA.

The scattering and absorption of light by small particles has been the object of extensive research, and there are several specialized monographs on the subject [16,17]. A flurry of activity is currently focused on the study of the fundamental mechanisms that make the optical, electronic, physical and chemical properties of these new materials so different from those of their bulk counterparts [18,19]. It is also a result of the desire to control the size, shape, structure and morphology of the nanostructures produced [20]. At present, several groups are exploring and developing optical characterization methods to capture the essential properties that allow one to define the fabricated nanostructures and promote applications. SERS, SERRS, SEIRA and reflection SEIRA (or SEIRRA) are part of this group of new nanostructure characterization techniques.

The objective of this brief optical discussion is to compile in one place the main results relevant to the phenomena of SERS and SEIRA: optical properties, scattering and absorption of light by metal particles, local fields, and reflecting surfaces. Notably, local field enhancements and light polarization at reflecting surfaces are explained classically and the molecule, a quantum object needed to observe SERS (examined in Chapter 1), will not appear until the end, when we give an example using reflection spectroscopy. It should also be emphasized that the enhanced electromagnetic fields operating in surface-enhanced spectroscopy (SES) are independent of the presence of a molecule for their manifestation. We will start the discussion with the propagation of electromagnetic radiation, optical properties of materials and dispersion, followed by the scattering and absorption by small particles, and ending with the example of reflection–absorption of a molecular thin film deposited on a reflecting surface.

2.1 ELECTRIC PERMITTIVITY AND REFRACTIVE INDEX

The effect of matter on electric fields is termed electric polarization P (C m^{-2}), while the effect on magnetic fields is termed magnetization M. Since the materials to be discussed are non-magnetic ($\mu = 1$), the effect of magnetization can be neglected. Experimentally, the polarization effect was related early to the difference between the speed of light in vacuum and its speed in any other medium. The ratio of the speed of light in vacuum to the speed of light in another medium is defined as the *refractive index*, η (11). For gases, the refractive index was found to be closely approximated by the square root of the relative permittivity:

$\eta = \sqrt{\varepsilon}$. The relative permittivity is a dimensionless quantity defined as the ratio $\varepsilon = \epsilon/\varepsilon_0$. The permittivity of the vacuum is a constant, $\varepsilon_0 = 8.854 \times 10^{-12}$ F m^{-1}. However, the values of the refractive index for liquids and solids deviate considerably from the $\sqrt{\varepsilon}$ rule. The refractive index and the relative permittivity are frequency dependent, and this dependence is termed dispersion. At low frequencies below 10^6 Hz (hertz $= \mathrm{s}^{-1}$) the relative permittivity of solids is in the 2–20 range. In the optical frequency range (visible, infrared), where there is a significant electronic response, the relative permittivity is in the 1–10 range. The limiting values of the relative permittivity at low frequencies or static permittivity (static dielectric constant) are tabulated in textbooks and handbooks, and for a given substance is the ratio of the capacity of a condenser with that substance as the dielectric medium to the capacity of the same condenser with a vacuum as the dielectric medium. The dielectric constant is a function of temperature and frequency at which the alternating electric field varies. For instance, water has a static relative permittivity of 88 at 4°C, giving a value >9 for its square root. This high value of ε is mainly due to dipolar polarization contributions. However, at high frequencies only the electronic polarization will contribute to relative permittivity and the frequency-dependent value will be lowered. For example, the value measured using yellow light for water is $\varepsilon = 1.77$, giving a value of 1.33 for the refractive index of water. A comprehensive discussion of the dispersion of water can be found in Jackson's book [12]. For SERS and SEIRA applications, we are only interested in the relative permittivity at optical frequencies.

The above discussion makes it is clear that explaining this *frequency-dependent dielectric function* (or refractive index) is a challenging proposition, and this has been a central problem in optics. As will be seen, the theory requires that the optical properties of a material be represented by a complex dielectric constant, $\varepsilon = \varepsilon_1 + i\varepsilon_2$, with its corresponding complex refractive index, $\eta = n + i\kappa$. Here again, the refractive index is the square root of the dielectric constant: $\varepsilon = \eta^2$. Both properties are functions of frequency with dispersive properties [11,12,16]. Many techniques have been developed to measure the real and imaginary part of the refractive index across the electromagnetic spectrum, and more than 30 of them can be listed [21,22]. The optical properties of solids are commonly measured using reflectivity, and one of the most common methods found in textbooks and commercial spectroscopic software is the Kramers–Kronig analysis, which involves measurements of reflectivity over a wide spectral range. For SERS and SEIRA applications, we profit from the extensive database created by many groups for the real, $n(\omega)$, and imaginary, $\kappa(\omega)$, parts of the refractive index from the ultraviolet to the infrared region of the electromagnetic spectrum (see, for instance, www.astro.spbu.ru/JPDOC/entry.html) [23]. A relevant discussion of the dielectric function of solids, in the context of infrared spectroscopy, can be found in the book by Decius and Hexter [24], and a complete collection of the optical properties of materials is now commercially available.

Relating macroscopic permittivity to atomic or molecular structure is one of the main objectives of optics. In practice, molecules and nanoparticles can respond to an external time-dependent electric field by changing their volume (polarizability) and producing an oscillating induced dipole, $p = \alpha E$, where α is the polarizability of the molecule or of the metal particle. The polarizability is a response function of the molecule or the particle and is represented by a tensor in three-dimensional space. The special case of a diagonal matrix is used to define the scalar polarizability as the trace $\alpha = \frac{1}{3}(\alpha_{xx} + \alpha_{yy} + \alpha_{zz})$. Finding a way to evaluate this response function has been a central problem in physics, providing the bridge between the macroscopic Maxwell theory and the structure of the material. In the nineteenth century, Mossotti (1850) and Clausius (1879) established that the molecular polarizability could be directly related and evaluated using the relative permittivity [dielectric function $\varepsilon(\omega)$] of the bulk material. The relation is known as the Clausius–Mossotti equation [12,25]:

$$\alpha_{\text{mol}} = \frac{3}{N}\left(\frac{\varepsilon - 1}{\varepsilon + 2}\right)$$

where N is the particle density and ε is the relative dielectric constant at a fixed frequency. The same relationship given in terms of the refractive index was discovered independently by Lorentz (1880) and Lorenz (1881), and is known as the Lorentz–Lorenz equation:

$$\alpha = \frac{3}{N}\frac{n^2 - 1}{n^2 + 2}.$$

These equation relate the macroscopic–optical properties of the medium to the number and properties of the scattering particles. Polarizability is an atomistic property, and volume density describes the macroscopic electric polarization, P. Notably, the factor in the Clausius–Mossotti equation:

$$g = \left(\frac{\varepsilon - 1}{\varepsilon + 2}\right)$$

will become central to the electromagnetic enhancement mechanism (in SERS) when experimental conditions cause the denominator to approach zero.

After the qualitative description of the macroscopic dielectric permittivity and the molecular property – polarizability – Maxwell's theory is needed to give physical meaning to these results and to explain the dispersion further. A discussion of the electrical nature (electrons and nuclei) of matter is also necessary to estimate polarization, P, and conductivity.

2.2 PROPAGATION OF ELECTROMAGNETIC WAVES AND THE OPTICAL PROPERTIES OF MATERIALS

The subject matter here is not the origin of the electromagnetic waves [11] but the propagation of electromagnetic radiation in solids, a refractive medium. Solids are broadly classified according to the value of the conductivity within: dielectrics $\sigma = 0$ and conductors or metals $\sigma \neq 0$. A perfect conductor is a material with $\sigma = \infty$. The optics of semiconductors is poorly described using the classical approach, and a quantum-theoretical treatment beyond the scope of the present discussion is required. The Maxwell equations (ME) view matter as a continuum. They are simplified in their application to solids by the assumptions that the solid is *isotropic*, *at rest* and free of *static charges*. In this macroscopic theory, space averages of all field quantities are performed over large distances and long time intervals compared with the characteristic atomic and molecular sizes and periods. The properties of the medium are found in the relationship between field vectors and are called the *materials equations* or *constitutive relations*. The field vectors in the presence of matter are the electric displacement, $D\,(\mathrm{C\,m^{-2}})$, magnetic flux density, B ($\mathrm{Wb\,m^{-2}}$ or T) and current density, $J\,(\mathrm{A\,m^{-2}})$. These are related to the electric field, E ($\mathrm{V\,m^{-1}}$), and magnetic field intensity, H ($\mathrm{A\,m^{-1}}$), through the *constitutive relations*, which are needed to solve the field vector ME:

$$D = \varepsilon_0(1+\chi)E = \varepsilon_0\varepsilon(\omega)E \tag{2.1}$$

$$B = \mu H \tag{2.2}$$

$$P = \varepsilon_0 \chi E \tag{2.3}$$

$$J = \sigma E. \tag{2.4}$$

These are linear relationships, and thereby, by definition, we are dealing with linear optics. Here, $\sigma(\Omega^{-1}\,\mathrm{m}^{-1})$ is the conductivity, μ is the magnetic permeability, $\varepsilon = 1 + \chi$ (dimensionless) is the relative permittivity and χ (dimensionless) is the electric susceptibility. The electric field in the ultraviolet and visible region of the spectrum displaces the electrons from their rest positions, creating a dipole density that marks the difference between $\boldsymbol{E}$ and $\boldsymbol{D}$. The electric displacement is written as $\boldsymbol{D} = \varepsilon_0 \boldsymbol{E} + \boldsymbol{P}$, where $\boldsymbol{P}\,(\mathrm{C\,m^{-2}})$ is the dipole polarization [for a discussion on the physical meaning of $\boldsymbol{P}$ see, for instance, Sommerfeld [26]]. For dielectrics and metals, the complex tensor $\varepsilon(\omega)$ in Equation (2.1) formally describes the dispersive properties of dielectric and conducting solids. The predominant property depends on the relative magnitudes of the period of the field ($T = 2\pi/\omega$) and the dielectric relaxation time $\tau = \varepsilon\varepsilon_0/\sigma$ (i.e. the time it takes for a charge imbalance to correct itself). When $T \gg \tau$, charges follow the field and the conductivity contribution to $\varepsilon_0\varepsilon(\omega)$ dominates as is the case in metals. For $T \ll \tau$ (slow relaxation), the dielectric polarization effect is predominant in $\varepsilon(\omega)$. It can be seen that for a perfect conductor the relaxation time is zero and the conductivity contribution prevails. Equations (2.1)–(2.4) are dependent on the medium under investigation, but will be assumed to be independent of field, position and direction [12].

Light is considered to be an oscillating electromagnetic wave composed of electric and magnetic fields, $\boldsymbol{E}$ and $\boldsymbol{H}$, perpendicular to one another travelling in the direction z perpendicular to the $\boldsymbol{EH}$ plane (consequently the name *plane wave*). This classical model, Maxwell's view of light as an electromagnetic wave, provides the vocabulary for most spectroscopic discussions. The electric field $\boldsymbol{E}$ oscillates sinusoidally in space and time, and the distance between successive peaks in the amplitude of $\boldsymbol{E}$ is termed the wavelength, λ. The direction of the electric field $\boldsymbol{E}$ in space defines its *polarization* (not to be confused with the electric polarization $\boldsymbol{P}$ in materials). Therefore, light propagating along the z-axis can be linearly polarized along the x-axis or linearly polarized along the y-axis. There is no component of $\boldsymbol{E}$ in the direction of propagation. The wavelength, λ, of the electromagnetic radiation is used to separate the electromagnetic spectrum into spectral regions. The ultraviolet (10 nm $< \lambda <$ 380 nm), visible (380 nm $< \lambda <$ 780 nm), infrared (0.8 μm $< \lambda <$ 1000 μm), including the near-infrared (0.8 μm $< \lambda <$ 2.5 μm) mid-infrared (2.5 μm $< \lambda <$ 50 μm) and far-infrared (50 μm $< \lambda <$ 1000 μm) are regions of interest for Raman and infrared measurements. The wavenumber unit is most commonly used in infrared and Raman spectroscopy; it gives the number of electromagnetic waves that fit in 1 cm. For instance, 500 $\mathrm{cm^{-1}}$

is equivalent to a $\lambda = 20000$ nm. Since in 1 cm there are 10^7 nm, we can fit $10^7/(2 \times 10^4) = 500$ wavelengths in 1 cm, each one of 2×10^4 nm. Propagating waves (solution of the Maxwell's equations) transport energy from one point to another. Maxwell's equations in point form and SI units are as follows:

$$\nabla \cdot D = 0 \qquad \text{Gauss's electric law}$$
$$\nabla \cdot B = 0 \qquad \text{Gauss's magnetic law}$$
$$\nabla \times E = -\frac{\partial B}{\partial t} \qquad \text{Faraday's law}$$
$$\nabla \times H = \frac{\partial D}{\partial t} + J_c \qquad \text{Ampère's law.}$$

Ampère's law contains the displacement current density $\partial D/\partial t = J_\mathrm{d}$ ($\mathrm{A\,m^{-2}}$), a term introduced by Maxwell that is crucial for high-frequency fields. It completes the set of equations that predict that light is an electromagnetic wave. The ratio between these two currents is $J_\mathrm{c}/J_\mathrm{d} = \sigma/\omega\,\varepsilon_0\varepsilon$. In fact, materials are commonly classified according to this ratio. Therefore, for good conductors $\sigma/\omega\varepsilon_0\varepsilon \gg 1$ and for a dielectric $\sigma/\omega\varepsilon_0\varepsilon \ll 1$. For example, for copper at 1 MHz, the ratio is ca 10^{12}, and this enormous difference is found for all 'conductors', providing a rationale for neglecting the displacement currents when studying light propagation in conductors. At the same time, this ratio is negligible for insulators. For example, at 1 MHz the ratio for Teflon is 2.6×10^{-4}, and in this case conduction currents can be neglected. This explains the fact that in optics, dielectrics and conductors are treated separately. Well-known manipulations of Maxwell's, and the constitutive equations, yield the wave or Helmholtz equations. The equations are composed of three scalar differential equations in terms of the component of the vector. The vector function E satisfies the vector wave equation when E is a solution of the scalar wave equation (see Bohren and Huffman [16]). The case of linearly polarized light can be written in the general form

$$\nabla^2 E_x + k^2 E_x = 0 \tag{2.5}$$

where $k\,(k^2 = \omega^2\varepsilon_0\,\varepsilon\,\mu_0\mu)$, is a complex number, referred to as the propagation constant of the medium and we should distinguish four cases:

1. In vacuum, $k = \omega/c = 2\pi/\lambda_0$ is merely the wavevector in wavenumbers, where c is the speed of light ($c = 1/\sqrt{\mu_0\varepsilon_0}$). This

represents the wave traveling, without distortions, where there is no dispersion.

2. In dielectric media, $k = \sqrt{\mu_0 \mu \times \varepsilon_0 \varepsilon} \times \omega$, and for non-magnetic media $\mu = 1$. Therefore, $k = \sqrt{\mu_0\, \varepsilon_0 \varepsilon} \times \omega = (\omega/c)\sqrt{\varepsilon} = (\omega/c)\eta$. This medium shows a dispersion proportional to the refractive index η.
3. For conductors, and for nonmagnetic materials, $k = \sqrt{i\omega \times \mu_0 \times \sigma}$. In this case, the real and imaginary part of k are equal, $k = \alpha + i\beta = (1+i)\sqrt{\omega\sigma\, \mu_0/2}$, and $\alpha = \beta = \sqrt{\omega\sigma\, \mu_0/2}$.
4. The general case of the propagation constant is $k = \sqrt{i\omega\, \mu_0 \mu\, (\sigma + i\omega\, \varepsilon_0 \varepsilon)}$. For nonmagnetic materials we can write

$$k = \frac{\omega}{c}\sqrt{\varepsilon\left(1 - i\frac{\sigma}{\varepsilon_0 \varepsilon\, \omega}\right)}.$$

The latter k reveals an expression of the dielectric function containing the contributions from bound (dielectric) and free electrons (conductor) in the material: $\varepsilon\,(\omega) = \varepsilon^{\text{bound}} - i\sigma/\varepsilon_0\omega$, or $\varepsilon\,(\omega) = \varepsilon^{\text{bound}}\,(\omega) + \varepsilon^{\text{free}}\,(\omega)$. To examine further the wave distortions in the medium, a physical model of dispersion is needed, so practical equations for the dielectric function can be derived. First, let us introduce an expression for the linearly polarized wave.

The propagation constant enters in the solutions of Equation (2.5), which are the same as those for a stretched string or a sound wave. Of the four possible solutions, we select a solution of the form $E_x = E_{0x}\exp[i(kz - \omega t)]$, with complex numbers written as $a' + ia''$. The latter is a linearly polarized wave with $E_y = E_z = 0$. Lets us examine the solution for case 2. Here, z is the direction of propagation, and the wavevector k contains information about the medium: $k = (\omega/c)\eta = (\omega/c)\,(n + i\kappa)$, where the factor η is the complex refractive index of the medium. The real and imaginary parts of the refractive index, η, are the experimentally collected optical 'constants' of materials, index of refraction (n) and extinction coefficient, or extinction index (κ). We can now rewrite the solution to express the physical situation of the propagating wave:

$$E_x = E_{0x}\exp\left(-\frac{\omega}{c}\kappa z\right) \times \exp\left[i\left(\frac{\omega}{c}n - \omega t\right)\right].$$

It can be seen that the first exponential represents the damping of the wave proportional to the extinction coefficient κ and the penetration depth z. Since the energy in the wave is proportional to $|E^2|$, the energy diminishes with distance as $\exp[-2(\omega/c)\kappa z] = \exp(-az)$, where a represents the *coefficient of absorption* of the medium. The penetration depth of any material is defined as the distance at which the amplitude of an electromagnetic wave drops to e^{-1} or 37%.

2.2.1 Frequency Dispersion in Solids

In solids, there is a difference between the applied field and the local field that may provide the most interesting optical phenomena. Here, for the time being, we assume them to be equal to keep the discussion in simple terms. The physical model for dielectrics uses the atomistic theory (electrons and nuclei) to explain the dispersion of the dielectric function. However, in condensed matter, there are many different types of electrons (oscillators) that contribute independently to the optical property. The approach gives a first approximation, but only quantum mechanics will give results that are close to observation. Nevertheless, the classical model provides the basis and the vocabulary for the interpretation of observed spectra, and the plan for the reminder of this chapter is to look into the properties of the oscillator, discuss the absorption by metal particles and end with the reflection–absorption of molecules on smooth metal surfaces.

First, the contribution of bound electrons to the dielectric function ($\varepsilon^{\text{bound}}$) is derived. The classical Lorentz oscillator model postulates that Hooke's law can describe the force binding the electron to the nucleus of an atom, $F(x) = -kx$, where x is the displacement from equilibrium. If the Lorentz system comes into contact with an electric field, then the electron will simply be displaced from equilibrium (21). The oscillating electric field of the electromagnetic wave will set the electron into harmonic motion. Hence, the dipole moment of each oscillator in the medium can be defined as $p = ex$. If the density of oscillators in the media is denoted N, then the polarization density, P, induced in the medium by the field may be defined as $P = Np = \varepsilon_0 \chi E = \varepsilon_0 (\varepsilon - 1) E$, where the magnetic field effects are ignored, $\mu = 1$. The equation of motion for Lorentz force oscillations, in an electric field in the x-direction, which takes *damping* effects into consideration, assumes the following form:

$$\frac{\partial^2 x}{\partial t^2} + \gamma \frac{\partial x}{\partial t} + {\omega_0}^2 x = -\frac{e}{m} E \tag{2.6}$$

where γ is the damping factor, ω_0 is the natural frequency of the oscillator, e is the electron charge and m is the electron mass [27]. Furthermore, assuming the electric field has the form $E_x = E_{0x}e^{-i\omega t}$, displacement follows the oscillations of the electric field:

$$x = x_0 e^{-i\omega t}, \quad \frac{\partial x}{\partial t} = -i\omega x_0 e^{-i\omega t}, \quad \frac{\partial^2 x}{\partial t^2} = -\omega^2 x_0 e^{-i\omega t}.$$

Substituting x and its derivatives into Equation (2.6) and solving for x_0:

$$x_0 = \frac{\frac{e}{m}}{\omega_0{}^2 - \omega^2 - i\gamma\omega} E. \tag{2.7}$$

The dipole moment and the polarizability α may now be defined as

$$p = ex_0 = \alpha E = \frac{\frac{e^2}{m}}{\omega_0{}^2 - \omega^2 - i\gamma\omega} E. \tag{2.8}$$

Since the polarization density is $P = Np$, taking the second derivatives and solving Equation (2.5) for k:

$$k^2 = \frac{\omega^2}{c^2}\left[1 + \frac{Ne^2}{\varepsilon_0 m} \times \frac{1}{(\omega_0{}^2 - \omega^2 - i\gamma\omega)}\right]. \tag{2.9}$$

A constant for each medium called the plasma frequency, $\omega_p{}^2$, is defined as $\omega_p{}^2 = \mathrm{N}e^2/\varepsilon_0 m$. We have now an expression for the dispersion of the dielectric function, $\varepsilon(\omega) = \eta^2$, and its relation to the polarizability:

$$\varepsilon(\omega) = \eta^2 = 1 + \frac{\omega_p{}^2}{\omega_0{}^2 - \omega^2 - i\gamma\omega} = 1 + N\alpha. \tag{2.10}$$

In order to account for each individual contribution, an *oscillator strength*, f_j, is introduced [24]. Then the dielectric function may be written as

$$\varepsilon'(\omega) + i\varepsilon''(\omega) = 1 + \sum_j f_j \frac{\omega_p{}^2}{\omega_j{}^2 - \omega^2 - i\gamma_j\omega}. \tag{2.11}$$

Extracting the real and the imaginary parts from Equation (2.10):

$$n^2 = 1 + \frac{\omega_p{}^2}{\omega_0{}^2 - \omega^2 - i\gamma\omega} \times \frac{\omega_0{}^2 - \omega^2 + i\gamma\omega}{\omega_0{}^2 - \omega^2 + i\gamma\omega} \tag{2.12}$$

$$n^2 = 1 + \frac{\omega_p{}^2(\omega_0{}^2 - \omega^2)}{(\omega_0{}^2 - \omega^2)^2 + \gamma^2\omega^2} + \frac{i\omega_p{}^2\gamma\omega}{(\omega_0{}^2 - \omega^2)^2 + \gamma^2\omega^2} = 1 + N(\alpha' + i\alpha''). \tag{2.13}$$

Hence, the real part of the complex dielectric function is found to be

$$n^2 - \kappa^2 = 1 + \frac{\omega_p{}^2(\omega_0{}^2 - \omega^2)}{(\omega_0{}^2 - \omega^2)^2 + \gamma^2\omega^2} = 1 + N\alpha' = \varepsilon_1(\omega) \tag{2.14}$$

and the imaginary part of the dielectric function is

$$i2n\kappa = \frac{i\omega_p{}^2\gamma\omega}{(\omega_0{}^2 - \omega^2)^2 + \gamma^2\omega^2} = iN\alpha'' = i\varepsilon_2(\omega). \tag{2.15}$$

The real and imaginary parts of the complex refractive index are given by

$$n = \left(\frac{1}{2}\{\varepsilon_1(\omega) + [\varepsilon_1{}^2(\omega) + \varepsilon_2{}^2(\omega)]^{\frac{1}{2}}\}\right)^{\frac{1}{2}} \tag{2.16}$$

$$\kappa = \left(\frac{1}{2}\{-\varepsilon_1(\omega) + [\varepsilon_1{}^2(\omega) + \varepsilon_2{}^2(\omega)]^{\frac{1}{2}}\}\right)^{\frac{1}{2}}. \tag{2.17}$$

In Equation (2.11), the damping constant, γ_j, is small compared with the corresponding frequency, ω_j. When in Equation (2.11) $\omega_j = \omega$, i.e. the external frequency reaches the value of one of the natural frequencies of the medium (particle, molecule), the system is said to be in resonance. The energy transfer or absorption of energy by the oscillator has a maximum at the resonant frequency, and the magnitude of the energy transfer is discussed in terms of a number defined for such forced oscillations as the *quality factor*, $Q = \omega_j/\gamma$. The relative intensity of the absorption is directly proportional to the magnitude of the quality factor. As can be seen from Equations (2.8) and (2.11), the imaginary part of the polarizability and the dielectric function have a singularity at the resonance frequency that is controlled only by the damping factor. The effect of the

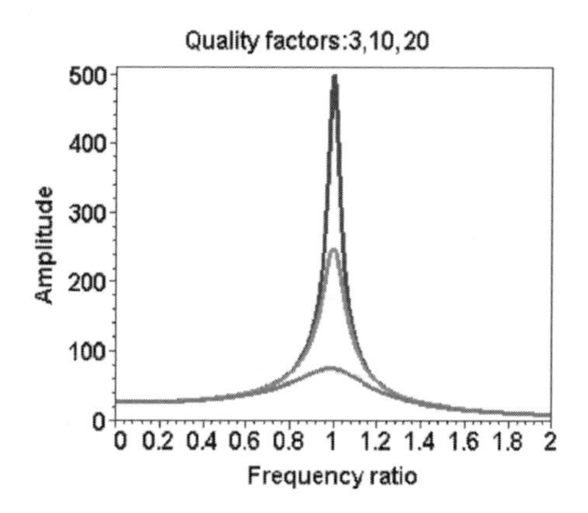

Figure 2.2 Amplitude vs frequency ratio illustrating the effect in the absorption spectrum of increasing quality factor Q from 3 (bottom) to 20

quality factor in the absorption spectrum is illustrated in Figure 2.2 using three values of Q, 3, 10 and 20. The maxima correspond to resonances, i.e. at $\omega/\omega_j = 1$.

It has been shown that the model of a single oscillator, or that of an ensemble of uncoupled oscillators, allows one to discuss the optical properties of dielectrics and semiconductors [27]. The amplitude of these driven oscillators is maximized in resonance, hindered only by damping, where a low damping corresponds to a high quality factor and vice versa. However, the amplitude strongly depends on the oscillator strength, a factor included in Equation (2.11), which is the strength of the coupling between the electromagnetic field and the oscillator, and in quantum mechanics is the square of the transition matrix element [24,28]. The resonances of oscillators allow for the description of different molecular resonances in the visible or infrared spectral region and also other resonances in solids, such as optical phonons, plasmons and excitons. The case of strong coupling between the electromagnetic wave and the polarization wave gives rise to the world of polaritons and quasiparticles.

2.2.2 Metals

Let us consider case 3, *k for conductors*. The existence of conductivity indicates that the conduction current, $J = \sigma E$, prevails over the displacement current, $\partial D/\partial t$. The simple solution of $k^2 = i\omega\mu_0\sigma$ is indeed a limited case for low frequencies. It is necessary to solve again the wave

Equation (2.5), introducing the dynamic expression of the current density, obtained by solving the equation of motion for unbound electrons [14], $J = (\sigma/1 - i\omega\tau)E$, where τ is the relaxation time of the transient current, which reduces to the static case $J = \sigma E$ when $\omega = 0$. The expression for the dielectric function for the free electrons is found to be given by

$$\varepsilon(\omega) = 1 - \frac{{\omega_p}^2}{\omega^2 + i\omega\tau^{-1}}. \tag{2.18}$$

For conductors, the penetration depth or *skin depth* is given by $\delta = \sqrt{2/\omega\sigma\mu_0}$. It can be seen that for good conductors the *skin depth* decreases rapidly with increasing frequency, $\delta = 6.52 \times 10^{-2}\sqrt{\lambda/c} = 3.77 \times 10^{-6}\sqrt{\lambda}$ with λ in m. For instance, for copper, $\sigma = 5.8 \times 10^7$ $\mathrm{S\,m^{-1}}$, the penetration depth at 6×10^{14} Hz (or 500 nm) is 2.7×10^{-9} m or 2.7 nm; however, at 10 GHz ($\lambda = 0.03$ m), $\delta = 6.5 \times 10^{-7}$ m or 650 nm.

In summary, classical theory provides an expression that accounts for the contribution of bound electrons and free electrons (case 4 for k) to the optical property of the material. Therefore, for semiconductors, the total expression for the dielectric function would be, [27]

$$\varepsilon(\omega) = \eta^2 = 1 - \frac{{\omega_p}^2}{\omega^2 + i\omega\tau^{-1}} + \frac{{\omega_p}^2}{{\omega_0}^2 - \omega^2 - i\gamma\omega}. \tag{2.19}$$

An example of a dielectric is a perfect semiconductor with a completely filled valence band and an empty conduction band.

In conclusion, the coupling of distinct oscillators in materials to electromagnetic radiation defines the optical properties of the material under investigation. For instance, in a semiconductor material, several forced oscillations can be observed that are classified as optical phonons, excitons and plasmons. For simplicity, we can say that if we excite an electron from the valence band to the conduction band by absorption of a photon, we simultaneously create a hole in the valence band. Such excitations of many electron systems are called 'excitons'. Similarly, the 'gas of free electrons' can be driven to perform collective oscillations that are known as *plasma oscillations* or *plasmons*. The latter play a key role in the explanation of SERS and SEIRA electromagnetic mechanisms when they are excited in nanoscopic metal particles.

2.3 SCATTERING AND ABSORPTION BY NANOSCOPIC PARTICLES

Nanometric metal particles can absorb light directly, whereas smooth metal surfaces may not. Surface plasmons can be excited in small metal particles or surface protrusions, but cannot be directly excited on flat metal surfaces (specific geometry is required). This fact is of central importance in the explanation of SERS (SEIRA) as we know it. A practical delimitation of the SERS effect is to give the name SERS only to observations on light absorbing nanostructures that can support surface plasmon resonances. First, we deal here with the absorption of light by nanometric particles, followed by the reflection of light on smooth surfaces, and its application in selective enhancement of vibrational frequencies (surface selection rules) by reflectivity and light polarization. Notably, in nanoscopic structures, a collective motion of conduction electrons, or surface 'plasmon', may become resonantly excited by visible or near-infrared light. Similarly, for SEIRA, an infrared counterpart to the surface plasmon is the surface phonon [10], which happens to have weaker damping and thus offers the advantage of stronger and sharper optical resonances for enhancement applications in the mid-infrared spectral region.

In a paper published in 1908, Mie [2] presented a rigorous solution to Maxwell's equations that describes the extinction spectra (extinction = scattering + absorption) of a plane monochromatic wave by homogeneous spheres of arbitrary size. A complete description can be found in two excellent monographs [16,17]. The subject has been of great interest, and over the years several reviews have been published on the linear optical properties of isolated metal particles of arbitrary shape, with diameters up to a few hundred nanometers [4,18,19,29–32]. Details on the history of the study of light absorbed and/or scattered by small particles can be found in Kerker's book, [17].

When a small spherical metallic nanoparticle (much smaller than the wavelength of the incident radiation) is irradiated by a plane monochromatic wave, the oscillating electric field causes the conduction electrons to oscillate coherently. These collective oscillations of conduction electrons are termed *particle plasmons*, *Mie plasmons*, *surface plasmons* or, within the dipole approximation in the Mie theory, *dipole particle plasmon resonance* (DPPR) [33], and should not be confused with other surface plasmon resonance (SPR) oscillations that can occur at a plane–metal–dielectric interface [19,34] The location of the oscillation frequency in the electromagnetic spectrum is determined by the dielectric function, shape and size of the metal particle. Gans [35,36] adapted Mie's theory to include scattering and absorption of prolate and oblate ellipsoids smaller than the wavelength of light. A review of computational work for different shapes can be found in the excellent work of Papavassiliou [29], where calculations of scattering and absorption of light by spheroidal, cubic and some rectangular parallelepiped particles of several materials, are illustrated. Creighton and Eadon [32] reported calculated visible absorption spectra (DPPR) of 10 nm diameter spherical particles for 52 elements in vacuum, and also immersed in water. Mulvaney [31] studied the effects of immersing metallic spheres in different media and coatings. The most recent reviews, by El-Sayed [18] and Kelly et al. [19], present results for the extinction of electromagnetic radiation by silver spheroids and non-spheroidal particles. Mie theory, although derived for a single sphere, also applies to a distribution of equivalent spheres randomly distributed and separated by distances that are larger than the wavelength of the exciting light. One of the earliest treatments of light interacting with a random distribution of metallic spheres is due to Maxwell-Garnett (MG) [1]. However, MG theory is phenomenological in nature, and is better discussed as part of effective medium theories that are used in the approach to SEIRA [37]. The MG model was developed with the help of the Lorentz–Lorenz equation (includes the Lorentz field interactions between particles), and works particularly well when the islands are well separated. Its first application was to explain colors observed in discontinuous metal films, where it is assumed that distinct inclusions exist in a host, and particles are modeled as a set of spheres in a matrix, making up a film. Typical absorption spectra due to DPPR of a distribution of silver particles in a vacuum evaporated metal island film are shown in Figure 2.3, together with the calculated absorption for a silver sphere and a silver prolate of 3:1 aspect ratio. The width of the absorption of silver particles is narrow, indicating only its dependence on particle size and shape. In contrast, silver films show broad absorption bands in correspondence with a broad distribution of particle shapes and sizes. Figure 2.3 also illustrates how plasmon frequencies for the same metal strongly depend upon particle shape. The absorption of the Ag sphere has a single band below 400 nm. The prolate ellipsoid shows two bands, one below 400 nm and the other, more intense band, at 494 nm.

Nanoparticles and nanowires are now commonly used in SERS applications. The differences in their plasmon absorption spectra are illustrated in Figure 2.4 by gold nanoparticles and silver nanowires. The plasmon absorption is shown increasing during the synthesis of gold nanoparticles in solution. Since the metal particles in solution are spherical, single plasmon absorption is observed. However, the nanowires give

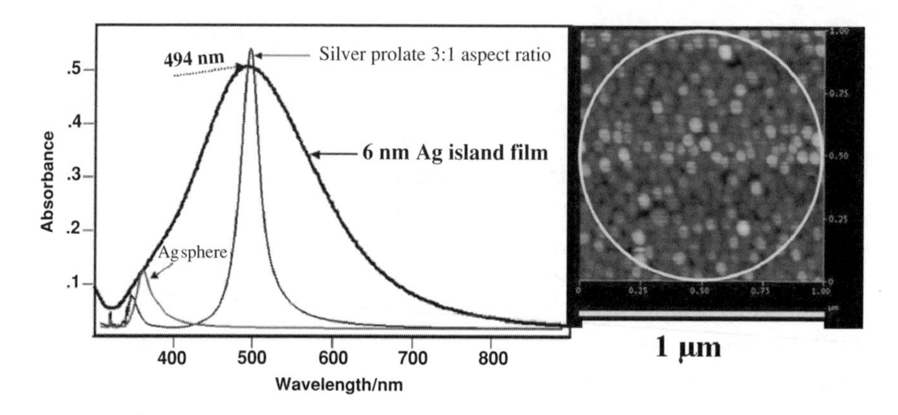

Figure 2.3 Experimental plasmon absorption and atomic force microscopy image of a 6 nm mass thickness silver evaporated film. The computed plasmon absorption spectra of a spherical Ag particle and a prolate Ag nanoparticle with 3 : 1 aspect ratio are also shown

rise to a tranverse oscillation (high frequency at 383 or 358 nm) and a longitudinal resonance in the near infrared region. Examples of images corresponding to gold nanoparticles and silver nanowires are shown in Figure 2.5. As was mentioned at the beginning of this section, computation of the plasmon absorption and scattering for small particles of different size is well documented in the literature, and a brief account of this work is given in the next section.

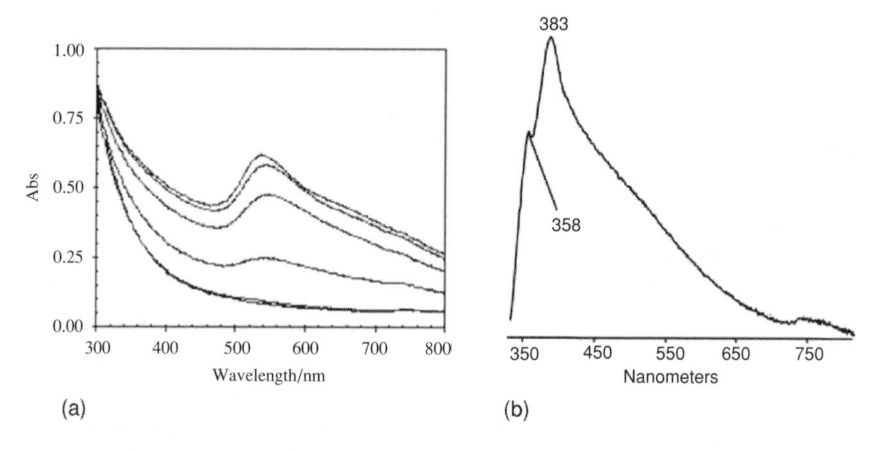

Figure 2.4 Plasmon absorption spectra of nanoparticles and nanowires. (a) Plasmon absorption of gold nanoparticles and (b) spectrum of silver nanowires

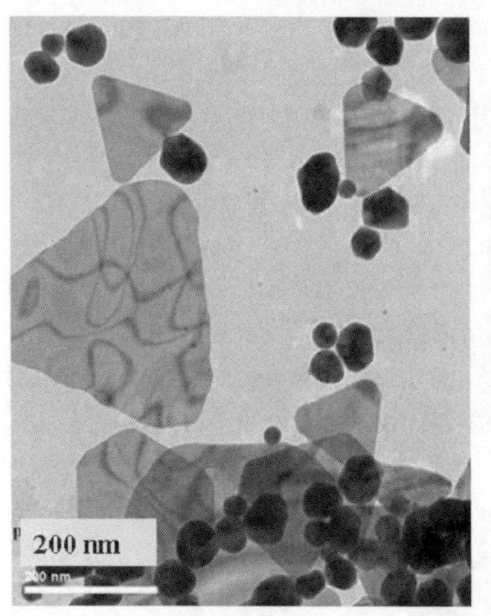

Figure 2.5 HTEM of gold nanoparticles and silver nanowires fabricated in our laboratory

2.3.1 Scattering and Absorption Computations

The computational task of calculating scattering and absorption cross-sections is facilitated if one considers particles that are small compared with the wavelength of light used in the experiment. This is called the electrostatic approximation, or Rayleigh theory of scattering. The approximate expressions correspond to truncated terms in the complete Mie treatment, and are widely used as a source of physical insight into the light scattering and absorption by small particles. In electromagnetic theory, the easiest way to evaluate electromagnetic fields that are radiated is to work with the electric potential Φ and the magnetic potential, rather than using Maxwell's equations directly. The energy flow of electromagnetic energy per unit of area is given by the Poynting vector, defined as the cross-product of the electric and magnetic fields. However, the time average of this vector has a simple expression indicating that the intensity is proportional to the square of the electric field: $I = |E|^2/2Z$. The quantity I is called irradiance (power per unit area) and Z is the impedance of the medium. In vacuum, $Z_0 = \sqrt{\mu_0/\varepsilon_0} = 377\Omega$.

The model used here is that of an isolated sphere embedded in a dielectric, ε_m, under the influence of an external field $\boldsymbol{E}_0$, a uniform field along the z-axis produced by incident linearly polarized light. This model, for sufficiently small spheres, produces exactly the same cross-section as the that for the Hertzian dipole radiation [38]. The field $\boldsymbol{E}$ oscillates temporally ($e^{-i\omega t}$), but is static spatially. For a sphere located at the origin the coordinate system, the primary potential at the limit when $r \to \infty$ is $\Phi_{\text{out}} = -E_0 z = -E_0 r \cos\theta$, the original external potential. Under these conditions, the electric potential satisfies Laplace's equation, $\nabla^2 \Phi = 0$, and the electric field is given by $E = -\nabla \Phi$ (17). The solution to Laplace's equation is a typical problem in electrostatics and is discussed thoroughly in textbooks (see reference 12, p. 95). Therefore, for an isotropic sphere of charge $\rho = 0$, with $\varepsilon(\omega)$, placed in a medium with dielectric constant ε_m, where there is a static electric field E_0, the solutions for the potential outside the sphere (Φ_{out}), and the potential inside the sphere (Φ_{ins}) are, in spherical coordinates

$$\Phi_{\text{out}} = -E_0 r \cos\theta + \left[\frac{\varepsilon(\omega) - \varepsilon_m}{\varepsilon(\omega) + 2\varepsilon_m}\right] \frac{a^3}{r^3} E_0 r \cos\theta \tag{2.20}$$

and

$$\Phi_{\text{ins}} = -\frac{3\varepsilon_m}{\varepsilon(\omega) + 2\varepsilon_m} E_0 r \cos\theta. \tag{2.21}$$

The complete solution for these potentials can also be found in Stratton's book (reference 39, p. 201). At this point, the convenient factors, commonly used in SERS,

$$g_0 = \frac{\varepsilon(\omega) - 1}{\varepsilon(\omega) + 2}$$

in vacuum where $\varepsilon_m = 1$ and

$$g_m = \frac{\varepsilon(\omega) - \varepsilon_m}{\varepsilon(\omega) + 2\varepsilon_m}$$

in a medium with relative permittivity ε_m can be introduced. The scattering potential is given by $\Phi_{\text{sca}} = g_m(a^3/r^3)E_0 r \cos\theta$. Our interest is to use these potentials to find the scattering fields and compute the intensity of the scattered radiation at a distance far from the sphere (far field). As pointed out before, the quantity directly related to the radiation is the square of the electric field, that, by definition, is the gradient of the potential. Therefore, finding the scattered field outside the sphere allows one to obtain the intensity of the scattered wave that is angularly dependent, and the scattering cross-section can then be obtained by integrating the energy scattered by the particle in all directions. The scattering problem for an arbitrary sphere was solved exactly by Mie. However, in most practical applications the quasi-static approximation is used, limiting the diameter of the sphere to be much smaller than the wavelength of the incident radiation $a/\lambda < 0.05$. Finding the fields for a sphere with radius a, where r is any distance from the centre of the sphere in spherical coordinates, the total field inside the sphere is $-\nabla\Phi_{\text{ins}} = E_{\text{ins}}$:

$$E_{\text{ins}} = \frac{3\varepsilon_m}{\varepsilon(\omega) + 2\varepsilon_m} E_0 e_z. \tag{2.22}$$

The scattering field outside the sphere is (the incident component is not included)

$$\begin{aligned} E_{\text{sca}} &= -\nabla\Phi_{\text{sca}} \\ E_{\text{sca}} &= g_{\text{m}}\frac{a^3}{r^3} E_0(2\cos\theta e_r + \sin\theta e_\theta). \end{aligned} \tag{2.23}$$

This part of the total external field, the *scattering field*, E_{sca}, is identical with that of a field created by a dipole p at the center of the sphere:

$$p = 4\pi\varepsilon_m \frac{\varepsilon(\omega) - \varepsilon_m}{\varepsilon(\omega) + 2\varepsilon_m} a^3 E_0 = \varepsilon_m 4\pi g_m a^3 E_0. \tag{2.24}$$

The field of the dipole is cylindrically symmetrical about the axis and the total field is

$$E_{\text{dip}} = \frac{p}{4\pi\varepsilon_m r^3}(2\cos\theta e_r + \sin\theta e_\theta). \tag{2.25}$$

From the induced dipole Equation (2.24), the expression for the polarizability can be extracted:

$$\alpha = 4\pi a^3 \frac{\varepsilon(\omega) - \varepsilon_m}{\varepsilon(\omega) + 2\varepsilon_m} = 4\pi a^3 g_m. \tag{2.26}$$

It can be seen that the polarizability depends on the volume of the sphere and has the units of volume. Incidentally, since our interest is in high-frequency optical fields, the polarization of the particle (polarizability α) will be entirely due to electron displacements.

The integrated power scattered from the sphere will also contain the incident field. "The *scattering cross-section* of the sphere is defined as the ratio of the total scattered energy per second to the energy density of the incident wave" (reference 39, p. 569). The scattering cross-section, in the far field (large r), for a small sphere, that is identical with that of the Hertzian dipole, is given by

$$\sigma_{\mathrm{sca}} = C_{\mathrm{sca}} = \frac{k^4}{6\pi} |\alpha|^2 = \frac{8}{3}\pi k^4 a^6 \left| \frac{\varepsilon(\omega) - \varepsilon_m}{\varepsilon(\omega) + 2\varepsilon_m} \right|^2 = 128\pi^5 \frac{a^6}{3\lambda^4} g_m{}^2 \tag{2.27}$$

and has the dimensions of an area (reference 17, p. 37). The attenuation of a beam of light through a number of particles of equal size is associated with scattering and absorption, and the net result is the *extinction* of the incident beam. The extinction cross section is therefore the sum of the scattering and the absorption cross-sections:

$$\sigma_{\mathrm{ext}} = \sigma_{\mathrm{sca}} + \sigma_{\mathrm{abs}}. \tag{2.28}$$

The extinction by a single particle is discussed in several monographs; see, for example, reference 16, p. 69. The final results are more useful to us if given in terms of the complex polarizability of the sphere, α. The scattering cross-section σ_{sca} and the absorption cross-section σ_{abs} are given by (16)

$$\sigma_{\mathrm{sca}} = k\mathrm{Im}\,\{\alpha\} = 4\pi k a^3 \mathrm{Im} \left\{ \frac{\varepsilon(\omega) - \varepsilon_{\mathrm{m}}}{\varepsilon(\omega) + 2\varepsilon_{\mathrm{m}}} \right\}. \tag{2.29}$$

Equation [2.29] is a good approximation when the scattering is small compared with absorption, and one can write $\sigma_{\mathrm{abs}} = k\mathrm{Im}(\alpha)$. There is an additional electric field near the sphere that, although it has no energy flow outward, may be very large. A near-field scattering cross-section can be defined by integration on the surface of the sphere:

$$\sigma_{\mathrm{nf}} = \frac{\alpha^2}{6\pi} \left(\frac{3}{a^4} + \frac{k^2}{a^2} + k^4 \right) \tag{2.30}$$

2.3.2 Mie Computations

As pointed out before, the exact solution for the sphere exists and there is no need for the approximation. Chapter 4 in Bohren and Huffman's book [16] gives the mathematical basis of Mie theory, and the full Mie [2] expressions for σ_{sca} and σ_{ext}, given as a function of the scattering coefficients a_n and b_n, as reproduced here:

$$\sigma_{\text{sca}} = \frac{2\pi}{k^2} \sum_{n=1}^{\infty} (2n+1) \left(|a_n|^2 + |b_n|^2\right) \tag{2.31}$$

$$\sigma_{\text{ext}} = \frac{2\pi}{k^2} \sum_{n=1}^{\infty} (2n+1) \left[R_e \left(a_n + b_n\right)\right]. \tag{2.32}$$

Equations (2.27)–(2.30) are valid only for particles with radii between 3 and 25 nm (17,32); above 25 nm, quadrupole and other higher order terms in the summation in Equation (2.31) become important. Notably, the static approximation results for spheres of radius up to 20 nm are practically identical with the results of the full Mie calculations of the scattering cross-sections. The results shown in Figure 2.6 illustrate the

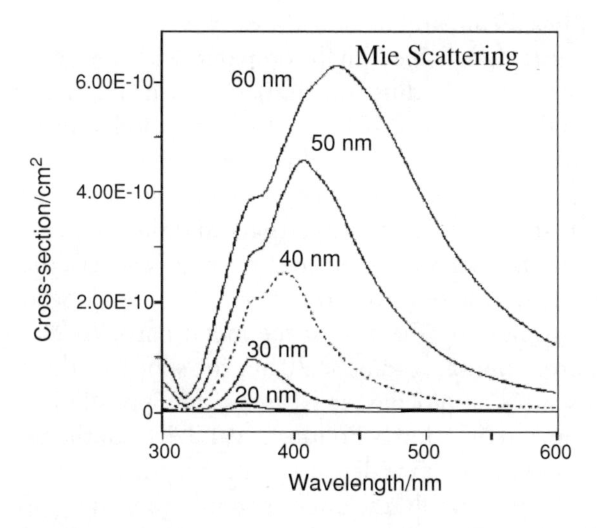

Figure 2.6 Mie scattering computational results illustrating the increase in scattering with increasing radii of silver spheres (the size effect)

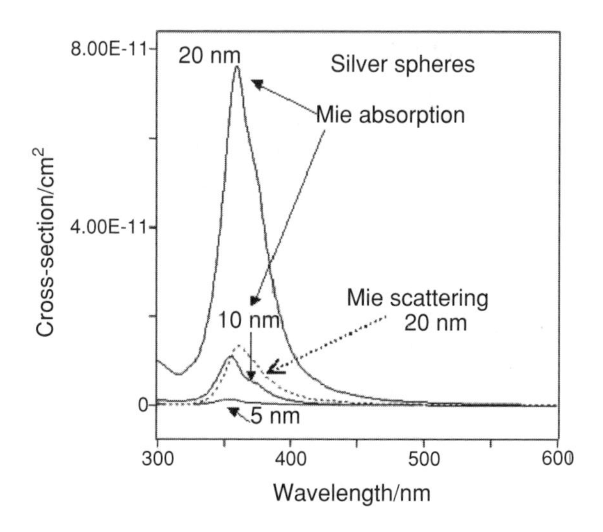

Figure 2.7 Results of Mie absorption and scattering computations for a silver sphere of 20 nm radius illustrating that for small particles (20 nm radius or less), scattering is indeed small compared with the absorption

increase in the Mie scattering with increased radii of silver spheres (the size effect).

Figure 2.7 shows the Mie absorption and scattering for a silver sphere of radius 20 nm and demonstrates that for small particles (radius $\leq$ 20 nm), scattering is indeed small compared with the absorption. However, for large particles (radius $>$ 30 nm), scattering is much stronger than absorption and Equation (2.29) is no longer a good approximation.

In summary, the computations illustrate three very important properties of the scattering and absorption of small metal particles. First, σ_{abs} and σ_{sca} peak at a well-defined wavelength and these maxima are caused by local plasmon resonance. The spectral region where the peak is observed depends on the optical properties of the metal. Second, σ_{abs} and σ_{sca} depend strongly on the size of the metal particle. Within the electrostatic approximation (radius $<$ 20 nm for spheres), the absorption is larger than scattering. For larger particles the opposite is true, and the electrostatic approximation is no longer valid. Third, the position of the peak for plasmon resonance depends on the refractive index of the surrounding medium. Notably, the intensity of plasmon absorption may also be affected by the surroundings of the particle. This effect is illustrated for gold spheres of 30 nm radius embedded in media with refractive indices of 1 and 1.5 in Figure 2.8.

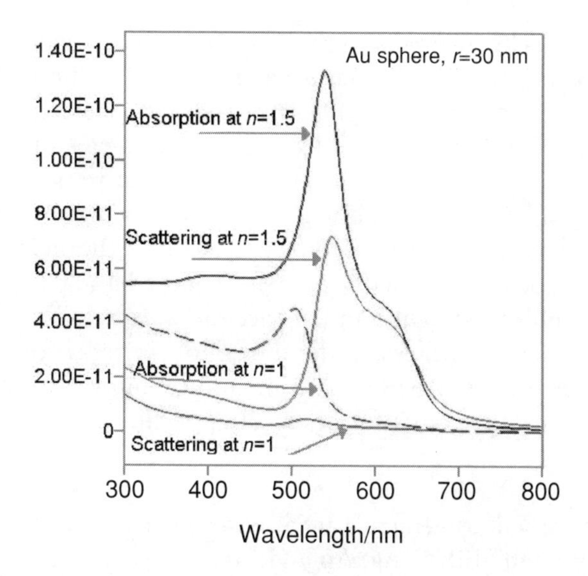

Figure 2.8 Results of computations illustrating the effect of the surroundings on the intensity of plasmon absorption of the nanoparticle. Computations for gold spheres of 30 nm radius embedded in media with refractive indices of 1 and 1.5

2.4 REFLECTION–ABSORPTION INFRARED SPECTROSCOPY ON SMOOTH METAL SURFACES

Whereas metal particles absorb electromagnetic radiation, smooth metal surfaces may not. Although collective oscillations of the electrons of smooth metal surfaces can also be excited, very special conditions are required, and they will give rise to a different physics and spectroscopy [40]. Notably, attenuated total reflection was first demonstrated as a method for excitation of surface plasmon by Otto in 1968 [41]. The title of his paper, 'Excitation of nonradiative surface plasma waves in silver by the method of frustrated total reflection', makes a clear reference to the nonradiative, surface electromagnetic waves which have interesting properties. In plasmon-assisted SERS, there is spontaneous emission of radiation by the induced dipole (radiative plasmons), and this emission grows with particle size. Normally, nonabsorbing smooth metal surfaces are reflecting surfaces. In our effort to delimit the SERS and SEIRA effects, it is profitable to discuss the reflection of light on surfaces that are not SERS or SEIRA active, and advance towards the scattering and reflection–absorption by molecules located near smooth metal surfaces. The treatment of the spectroscopic behavior of molecules on reflecting smooth metal surfaces belongs to the well-established field of *reflection spectroscopy* [42]. Correspondingly, there exist reflection Raman scattering spectroscopy [43] and reflection–absorption infrared spectroscopy [44] that take advantage of the selective probing that can be achieved with polarized light and molecular orientation on reflecting surfaces. The objective of this section is to facilitate the understanding of surface vibrational experiments by reviewing the basic properties of the reflection of polarized light and the spectral interpretation of absorbed molecules on metal surfaces, with an example in reflection–absorption infrared spectroscopy (RAIRS). The intention, simply, is to illustrate the following important points for practical application and spectral interpretation:

1. The most reflecting metals are also the metals most commonly used in SERS and SEIRA: Ag, Au and Cu.
2. The p-polarized light is stronger (positive interference) at the surface, and the surface polarization is more pronounced at higher incident angles, determining that RAIRS should be carried out at large angles ($\sim 80^\circ$) for highly reflecting metals.
3. The infrared absorption by molecular vibrations with a nonzero component of the dynamic dipole perpendicular to the surface will be strengthened, whereas the absorption by molecular vibrations with nonzero components of their dynamic dipole parallel to the surface will be weakened with respect to the free molecule.
4. The spectral interpretation may require that the local symmetry of the adsorption site be taken into account when determining the symmetry, or molecular orientation of the adsorbed species with respect to the reflecting surface, which must be considered in organized films.

It should be noted that a number of acronyms exist for the infrared absorption recorded using the external specular reflectance: RAIR, RAIRS, IRRS, IRRAS, FT-IRAS, RAS, GIR, IR-ERS, ERIR, FT-IRRAS and FTIR/RA. Here we used RAIRS.

2.4.1 Reflection Coefficients and Reflectance

In 1823, Fresnel proposed a derivation for the reflection coefficients of light waves that was later confirmed using Maxwell's equations. The

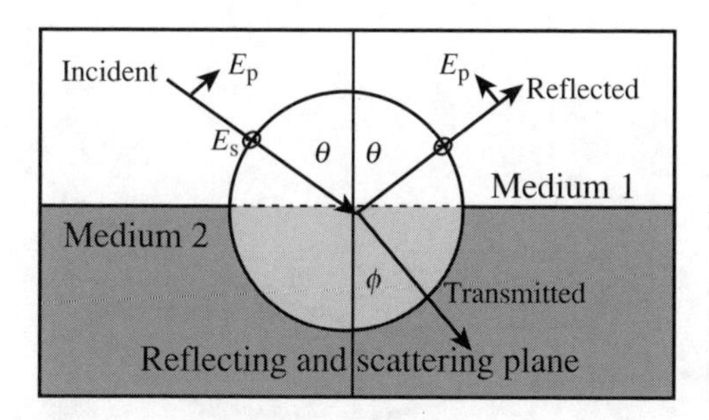

Figure 2.9 Definition of the reflecting and scattering plane for a plane wave incident at an arbitrary angle on a boundary plane

standard procedure is to resolve the electric field into two orthogonal components, or two polarizations. As is traditional in optics, the direction of the electric field is assumed to be the direction of polarization. The two most common cases of polarization used in reflection spectroscopy are the s and p polarized light. The definition is illustrated in Figure 2.9, where for a plane wave incident at an arbitrary angle on a boundary plane, the electric field vector can be parallel (p) or perpendicular (s – *senkrecht* in German) with respect to the plane of incidence. In optics, s-polarized light is called the transverse electric, or TE polarization, and the p-polarized light is termed the transverse magnetic, or TM polarization.

The Fresnel amplitude reflection coefficients for p- and s-polarizations are obtained by applying boundary conditions to the solutions of Maxwell's equations [11]: $k_i r = k_r r = k_t r$.

$$r_s = \frac{n_1 \cos\theta - \sqrt{n_2{}^2 - n_1{}^2 \sin^2\theta}}{n_1 \cos\theta + \sqrt{n_2{}^2 - n_1{}^2 \sin^2\theta}} \tag{2.33}$$

and

$$r_p = \frac{-n_2{}^2 \cos\theta + n_1\sqrt{n_2{}^2 - n_1{}^2 \sin^2\theta}}{n_2{}^2 \cos\theta + n_1\sqrt{n_2{}^2 - n_1{}^2 \sin^2\theta}} \tag{2.34}$$

where n_1 and n_2 are the refractive indices of the two media and θ is the angle of incidence [14]. The reflectance is defined as $R_s = |r_s|^2$ *and* $R_p = |r_p|^2$. Since the energy is proportional to the absolute square of the field amplitude, reflectance values are the energy ratio of the reflected to the incident light.

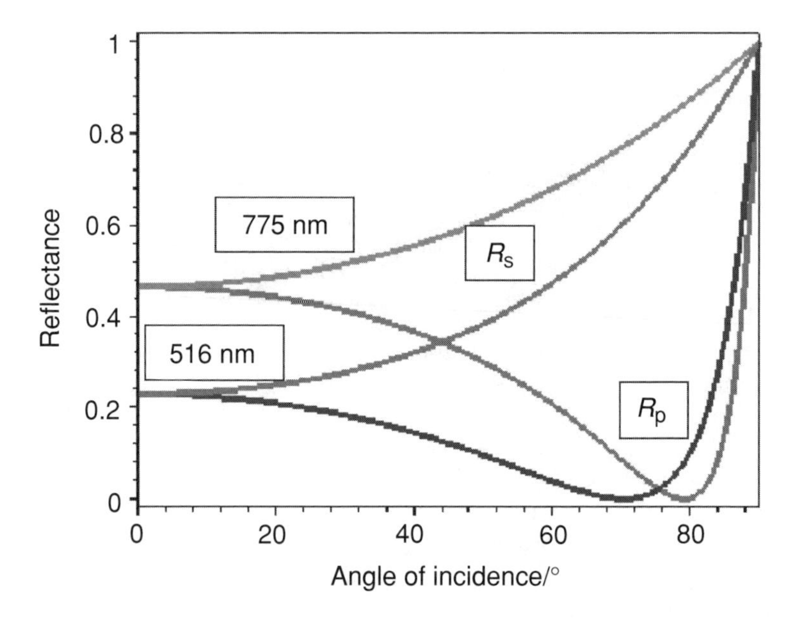

Figure 2.10 Calculated reflectance R_S and R_P plotted against the angle of incidence for two wavelengths, 516 and 775 nm, using the corresponding optical constants of silver

The reflectance R_s and R_p, against the angle of incidence, is plotted for two wavelengths using the optical properties of silver in Figure 2.10 to illustrate the peculiar behavior of p-polarized light versus s-polarized light. R_s increases steadily from the normal to the boundary plane, whereas R_p shows a decreasing behavior with a minimum at ca 60–80°. The minimum changes with the wavelength of the incident radiation, indicating that the optical 'constants' of the reflector are different for two wavelengths.

We now consider the interference of the incident and reflected waves when a plane monochromatic wave falls on a highly reflecting surface. The plane is $z = 0$, and the positive direction of z points into the medium where the wave is propagating.

The components of E incident are

$$E_x{}^i = -A_{\|}\cos\theta_{\mathrm{i}}\mathrm{e}^{-i\tau},\ E_y^i = A_{\perp}\mathrm{e}^{-i\tau},\ E_z^i = -A_{\|}\sin\theta_{\mathrm{i}}\mathrm{e}^{-i\tau} \tag{2.35}$$

where

$$\tau = \omega\left(t - \frac{x\sin\theta_i - z\cos\theta_i}{v}{}_1\right). \tag{2.36}$$

The reflected wave components are given by similar expressions:

$$E_x{}^i = R_{\|}\cos\theta_i e^{-i\tau}, \quad E_y{}^i = R_{\perp}e^{-i\tau}, \quad E_z{}^i = -R_{\|}\sin\theta_i e^{-i\tau}. \qquad (2.37)$$

However, it is important to remember that the amplitudes of reflected waves are related to the amplitudes of the incident wave by the Fresnel coefficients:

$$R_{\|} = r_{\rm p}A_{\rm p}, \quad R_{\perp} = r_{\rm s}A_{\perp}. \qquad (2.38)$$

From Equations (2.35)–(2.37), it can be found that

$$\begin{aligned} E_x &= \cos\theta_i\left(r_{\rm p} - 1\right)A_{\rm P} \\ E_y &= (1 + r_{\rm s})A_{\perp} \\ E_z &= -\sin\theta_i\left(1 + r_{\rm p}\right)A_{\rm P} \end{aligned} \qquad (2.39)$$

This set of Equations [2.39] can be used to calculate the primary field at the surface using the Fresnel coefficients. A plot of the square of the expressions given in Equations (2.39), for a silver surface at 516 nm, is shown in Figure 2.11, where z is the direction perpendicular to the

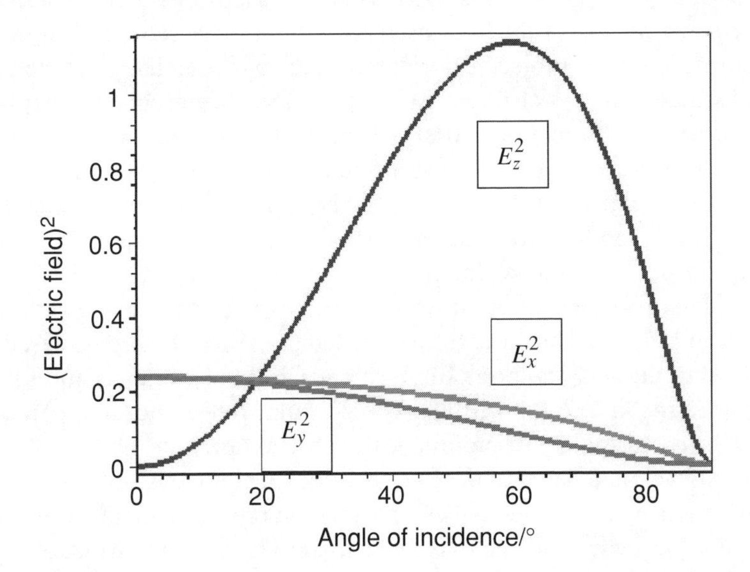

Figure 2.11 Plot of the square of electric field components for an incident wave at 516 nm on a silver surface as a function of the angle of incidence. The direction perpendicular to the surface is z

surface. It can be seen that at high angle of incidence, between 60 and 80°, the component of the electric field perpendicular to the surface dominates. The behavior observed in Figure 2.11 is also found for good reflectors in the near-and mid-infrared spectral regions, hence its application in RAIRS.

2.4.2 Reflection–Absorption Infrared Spectroscopy (RAIRS)

The selection rules for isolated molecules were explained in Section 1.7 in Chapter 1. It was shown that it is simple to see which vibrational modes are infrared or Raman active once the molecule is assigned to a symmetry point group. However, in solids and adsorbed species, the observed intensity of the allowed infrared or Raman modes may be modulated by the polarization of the incident electric field and a spatial molecular orientation. Allowed infrared modes of a given symmetry species will be seen with an absorption intensity proportional to the square of the scalar product, $E_j \cdot \mu' = |E| \left|\mu'\right| \cos\theta$, where θ is the angle between the direction of polarization of the vector E and the direction of the dynamic dipole μ' $(\partial\mu/\partial Q)$ for each vibration. Immediately one can see that for $\theta = 0°$ (parallel vectors), a maximum absorption would be seen, and for $\theta = 90°$, the absorption is zero. This is the origin of the local selection rules or *surface selection rules* used to explain observed infrared or Raman intensities for molecules oriented at the surface. The reflection and interference at metal surfaces help to provide a comprehensive explanation of the experimentally observed spectral intensities, and becomes a powerful tool for determining molecular orientation. The practical application of selection rules may simply be a case of *symmetry reduction*, as it is employed for molecules adsorbed on metal surfaces.

The corollary is that symmetry species of single crystals and adsorbed molecules of known orientation may be distinguished using polarized radiation (24,44). The basic optics needed to explain RAIRS is described in detail in the classical book by Born and Wolf [11]. Francis and Ellison [45] were the first to report the use of specular reflectance in a paper entitled 'Infrared spectra of monolayers on metal mirrors'. The subject was further developed for practical application of the basic optics in vibrational experiments on molecules adsorbed on reflecting metal surfaces by Greenler [46], who realized that the light at the reflecting surface is highly polarized, and that at the appropriate angle of incidence the p-polarized (TM wave) component of the electromagnetic wave is about three orders of magnitude larger than the parallel component (reference 11, p. 42).

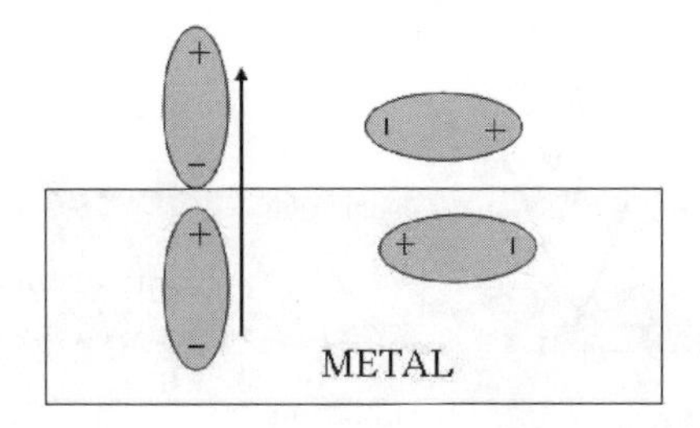

Figure 2.12 Image dipole picture representation showing that a dipole parallel to the reflecting surface is cancelled by its image, while the perpendicular dipole and its image are additive

Since the absorption of infrared radiation by a molecule is determined by the $E \cdot \mu'$ (scalar product), the term *surface selection rule* has been coined to emphasize the dependence of the observed intensities on the orientational properties of the molecule and light polarization at the surface. Therefore, the surface selection rules used in infrared and also in Raman spectroscopy, for reflecting surfaces, are a direct application of the electric field polarization at the metal surface and the molecular orientation.

Similar results are obtained with the image dipole picture, where a dipole parallel to the reflecting surface effectively is cancelled by its image, while the perpendicular dipole and its image add up as shown in Figure 2.12.

If the molecular orientation is known and the infrared reflecting surface is flat, E_z is the predominant polarization, and the use of this single polarized component is of great significance in distinguishing orientationally allowed modes. The technique is most valuable for extracting molecular orientation information for molecules adsorbed on flat metal surfaces [47].

2.4.3 RAIRS Example

RAIRS is currently associated with the single reflection technique. The multiple internal reflection technique, most commonly known as attenuated total reflection (ATR), and widely used in SEIRA experiments, is not discussed here. A detailed discussion of ATR and its many applications can be found in a book edited by Mirabella [48].

The example chosen to illustrate the use of RAIRS in molecular orientation studies is that of a thin solid film of naphthalic 1,8:4,5-dianhydride

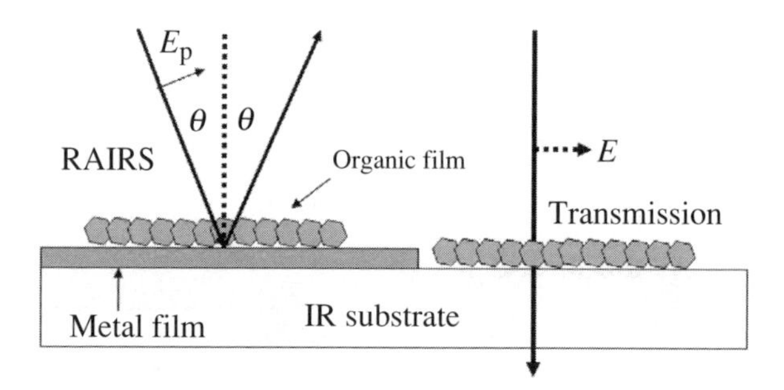

Figure 2.13 Experimental geometry for recording RAIRS and transmission FTIR spectra. A smooth metal film on the left-hand side gives rise to the reflection absorption experiment. The organic film on an infrared-transparent substrate (right-hand side) shows the transmission FTIR geometry

(NTCDA) [49]. The equilibrium geometry of NTCDA is planar, and belongs to the symmetry point group D_{2h}. There are a total of 66 fundamental vibrational normal modes divided into the following irreducible representations:

$$\Gamma = 12a_g + 4b_{1g} + 6b_{2g} + 11b_{3g} + 5a_u + \underline{\mathbf{11}\ \boldsymbol{b}_{1u}} + \underline{\mathbf{11}\ \boldsymbol{b}_{2u}} + \underline{\mathbf{6}\ \boldsymbol{b}_{3u}}$$

with the underlined bold-face symmetry species being infrared active. The molecule has a total of 22 in-plane normal modes and six out-of-plane vibrations. The experimental geometry is shown in Figure 2.13. A minimum of three infrared spectra are needed to extract qualitative information about the molecular orientation in organized films [50]:

- the reference spectrum for infrared-active fundamental normal modes of vibrations obtained in FTIR transmission experiments of the material dispersed in a KBr pellet or in solution;
- the FTIR transmission spectrum of the organized film deposited on a transparent IR substrate such as shown in Figure 2.15.
- the RAIRS spectrum of the same organized film deposited on a reflecting metal surface shown in Figure 2.16.

The experimental geometry is shown in Figure 2.13. The left-hand side corresponds to the reflection absorption experiment and the right-hand side is the transmission FTIR geometry.

The calculated spectra for each of the species of symmetry are shown in Figure 2.14. The 11 b_{1u} in-plane vibrations are polarized along the z

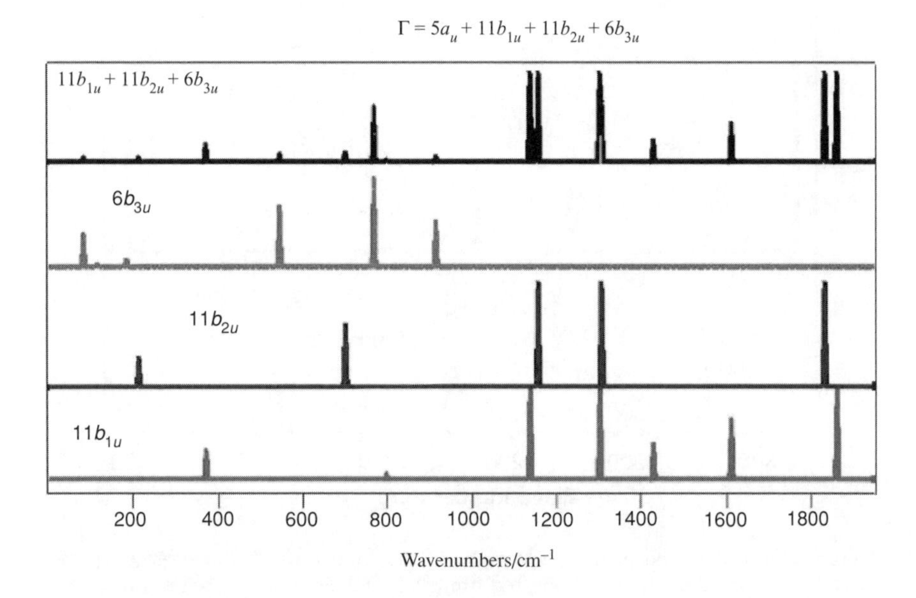

Figure 2.14 Calculated infrared spectra (top) and the spectrum for each of the infrared-allowed species of symmetry

molecular axis (the long molecular axis) and the 11 b_{2u} species along the short molecular axis (y). The six b_{3u} species are polarized along the x-axis. With perfectly aligned molecules and linearly polarized light, one should be able to obtain three distinct spectra, one for each of the symmetry species. For instance, a face-on molecular organization in the organic film deposited on a reflecting surface should show exclusively the six b_{3u} species in the RAIRS spectrum as shown in Figure 2.14.

An evaporated film of NTCDA on a transparent IR substrate shows the spectrum given in Figure 2.15 [49]. The spectrum is the sum of the $b_{1u}(z) + b_{3u}(x)$ species, indicating that the molecular alignment is edge-on, i.e. the molecule has its $z - x$ plane perpendicular to the substrate plane and therefore, $E \cdot \mu_z{}'$ and $E \cdot \mu_x'$ will give maximum absorption whereas $E \cdot \mu_y' = 0$ (b_{2u}) modes are silent. The transmission FTIR spectrum of the evaporated 70 nm mass thickness film on a KBr substrate is compared with the results of *ab initio* calculation [HF/6–311G(d,p) level of theory]. Given that the experiment gives a solid-state spectrum, the matching of the two spectra is very good and supports the extracted edge-on molecular orientation in the film.

Since the p-polarized (E perpendicular to the surface as shown in Figure 2.13.) component of the electromagnetic wave is about three orders

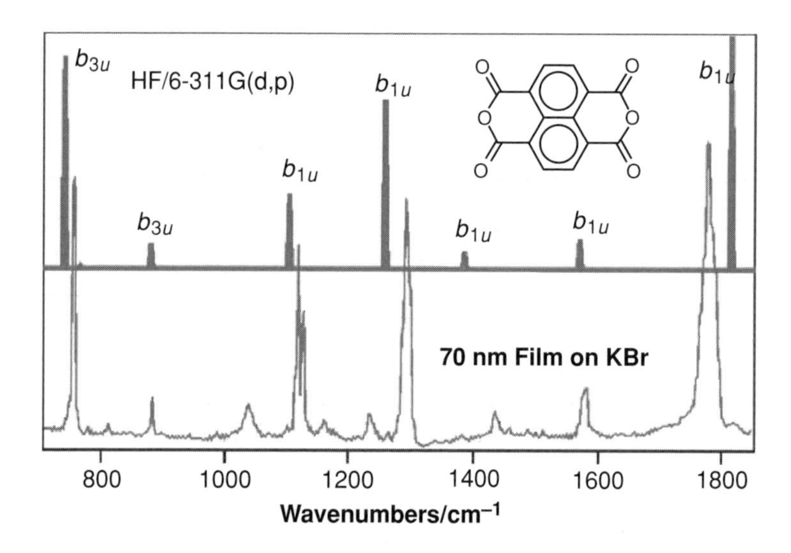

Figure 2.15 Calculated infrared spectrum of the sum of the $b_{1u}(z) + b_{3u}(x)$ species directly compared with an evaporated film of NTCDA on an infrared-transparent substrate. Reproduced from Aroca *et al.*, *Asian J. Phys.*, 1998, 7, 391–404, with permission from Asian Journal of Physics

of magnitude larger than the parallel component, the RAIRS spectrum shows strong relative intensity of the in-plane vibrations in the top spectrum given in Figure 2.16. The latter eliminates the presence of a dominant face-on molecular organization for the NTCDA film on smooth reflecting silver. The calculated vibrational spectrum shows that there

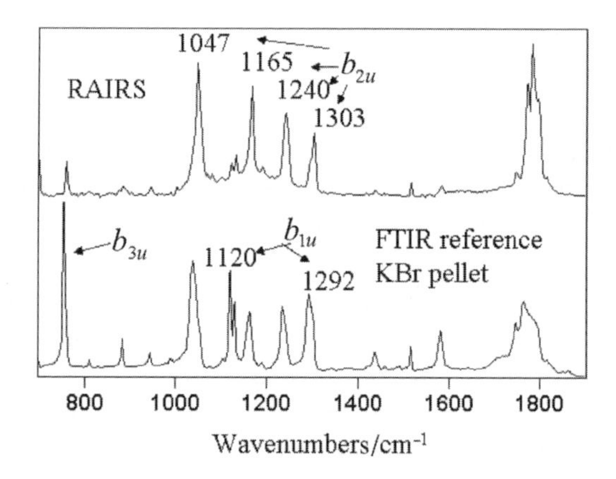

Figure 2.16 RAIRS spectrum of NTCDA film on smooth silver showing the strong relative intensity of the b_{2u} bands. The reference FTIR spectrum of NTCDA in a KBr pellet is included for comparison

are four b_{2u} (short y-axis) infrared-active vibrations in the wavenumber region from 1000 to 1350 cm^{-1}. The observed RAIRS spectrum does show four bands at 1047, 1165, 1240 and 1303 cm^{-1}, as can be seen in Figure 2.16, in agreement with the vibrational computations. At the same time, the $b_{1u}(z) + b_{3u}(x)$ species are observed with weak relative intensity, an indication of a preferential molecular orientation with the y-axis perpendicular to the reflecting surface, favoring the b_{2u} species. Therefore, the relative intensity of the b_{2u} bands is enhanced in the RAIRS spectrum, as expected for an edge-on, or preferentially edge-on, molecular orientation.

In conclusion, the spectral interpretation of both the RAIRS and transmission spectra of NTCDA films spectra formed on dielectric and metal surfaces are compatible with an edge-on molecular orientation as a property of the molecular organization of the organic film on the substrate plane. It is also confirmed that the relative intensities in the RAIRS spectra of NTCDA films on smooth reflecting silver surfaces were in full compliance with the surface selection rules for reflection–absorption infrared spectroscopy.

REFERENCES

[1] J.C. Maxwell-Garnett, Colours in metal glasses and in metallic films, *Philos. Trans. R. Soc. London, Ser. A*, 1904, **203**, 385–420.

[2] G. Mie, Considerations on the optics of turbid media, especially colloidal metal sols, *Ann. Phys.* 1908, **25**, 377–445.

[3] G. Schmid, (ed.), *Nanoparticles. From Theory to Applications.* Wiley-VCH, Weinheim, 2005.

[4] U. Kreibig and M. Vollmer, *Optical Properties of Metal Clusters.* Springer-Verlag, Berlin, 1995.

[5] D.L. Feldheim and C.A. Foss (eds), *Metal Nanoparticles. Synthesis, Characterization and Applications.* Marcel Dekker, New York, 2002.

[6] C.S.S.R. Kumar J. Hormes and C. Leuschner (eds), *Nanofabrication Towards Biomedical Applications.* Wiley-VCH, Weinheim, 2005.

[7] J. Li, X. Li, H.-J. Zhai and L.-S. Wang, Au_{20}: a tetrahedral cluster, *Science* 2003, **299**, 864–867.

[8] U. Kreibig, G. Bour, A. Hilger and M. Gartz, Optical properties of cluster-matter. Influences of interfaces, *Phys. Status Solidi A* 1999, **175**, 351–366.

[9] J. Zheng, C. Zhang and R.M. Dickson, Highly fluorescent, water-soluble, size-tunable gold quantum dots, *Phys. Rev. Lett.* 2004, **93**, 077402-1–077402-4.

[10] R. Hillenbrand, T. Taubner and F. Keilmann, Phonon-enhanced light-matter interactions at the nanometre scale, *Nature* 2002, **418**, 159–162.

[11] M. Born and E. Wolf, *Principles of Optics*, Pergamon Press, Oxford, 1975.

[12] D.J. Jackson, *Classical Electrodynamics*, John Wiley & Sons, Inc, New York, 1999.

[13] C.R. Paul and S.A. Nasar, *Introduction to Electromagnetic Fields*, McGraw-Hill, New York, 1987.
[14] G.R. Fowles, *Introduction to Modern Optics*, Holt, Rinehart and Winston, New York, 1975.
[15] D. Roy and J. Fendler, Reflection and absorption techniques for optical characterization of chemically assembled nanomaterials, *Adv. Mater.* 2004, **16**, 479–508.
[16] C.F. Bohren and D.R. Huffman, *Absorption and Scattering of Light by Small Particles*, John Wiley & Sons, Inc., New York, 1983.
[17] M. Kerker, *The Scattering of Light and Other Electromagnetic Radiation*, Academic Press, New York, 1969.
[18] M.A. El-Sayed, Small is different: shape-, size-, and composition-dependent properties of some colloidal semiconductor nanocrystals. *Acc. Chem. Res.* 2004, **37**, 326–333.
[19] K.L. Kelly, E. Coronado, L.L. Zhao and G.C. Schatz, The optical properties of metal nanoparticles: the influence of size, shape, and dielectric environment, *J. Phys. Chemi. B* 2003, **107**, 668–677.
[20] D.A. Fleming and M.E. Williams, Size-controlled synthesis of gold nanoparticles via high-temperature reduction, *Langmuir* 2004, **20**, 3021–3023.
[21] R.E. Hummel, *Electronic Properties of Materials*, Springer-Verlag, Berlin, 1985.
[22] W.A. Harrison, *Electronic Structure and the Properties of Solids*, Dover, New York, 1989.
[23] E.D. Palik (ed.), *Handbook of Optical Constants of Solids I, II and III*, Academic Press, New York, 1985, 1991, 1998.
[24] J.C. Decius and R.M. Hexter, *Molecular Vibrations in Crystals*, McGraw-Hill, New York, 1977.
[25] C.J.F. Böttcher, *Theory of Electric Polarization*, Elsevier, Amsterdam, 1973.
[26] A. Sommerfeld, *Optics*, Academic Press, New York, 1967.
[27] C.F. Klingshirn, *Semiconductor Optics*. Springer-Verlag, Berlin, 1997.
[28] E.B. Wilson Jr, J.C. Decius and P.C. Cross, *Molecular Vibrations; the Theory of Infrared and Raman Vibrational Spectra*, McGraw-Hill, New York, 1955.
[29] G.C. Papavassiliou, *Prog. Solid State Chem.* 1984, **12**, 185–271.
[30] J. Sukmanowski, J.-R. Viguié, François Arago, B. Nölting, and F. X. Royer, Light Absorption Enhancement by Nanoparticles, *J. Appl. Phys.* 2005, **97**, 104332–104337.
[31] P. Mulvaney, Surface plasmon spectroscopy of nanosized metal particles, *Langmuir* 1996, **12**, 788–800.
[32] J.A. Creighton and D.G. Eadon, Ultraviolet–visible absorption spectra of the colloidal metallic elements, *J. Chem. Soc., Faraday Trans.* 1991, **87**, 3881–3891.
[33] K.L. Kelly, E. Coronado, L.L. Zhao and G.C. Schatz, The optical properties of metal nanoparticles: the Influence of size, shape, and dielectric environment, *J. Phys. Chemi. B* 2003, **107**, 668–677.
[34] S. Kawata, M. Ohtsu and M. Irie, *Nano-optics*. Springer-Verlag, Berlin, 2002.
[35] R. Gans, *Ann. Phys.* 1912, **37**, 881.
[36] R. Gans, *Ann. Phys.* 1915, **47**, 270.
[37] D. Ross and R. Aroca, Effective medium theories in surface enhanced infrared spectroscopy: the pentacene example, *J. Chem. Phys.* 2002, **117**, 8095–8103.
[38] J.A. Kong, *Theory of Electromagnetic Waves*, John Wiley & Sons, Inc., New York, 1975.
[39] J.A. Stratton, *Electromagnetic Theory*, McGraw-Hill, New York, 1941.

[40] D.R. Tilley, *Basic Surface Plasmon Theory in Surface Plasmon–Polaritons*, IOP Short Meetings Series No. 9, Institute of Physics, IOP Publishing, London, 1987, pp. 1–24.
[41] A. Otto, Excitation of nonradiative surface plasma waves in silver by the method of frustrated total reflection, *Z. Phys.* 1968, **216**, 398–410.
[42] W.W. Wendlandt and H.G. Hecht, *Reflectance Spectroscopy*, John Wiley and Sons, Inc., New York, 1966.
[43] A. Campion, in J.T. Yates Jr and T.E. Madey (eds), *Vibrational Spectroscopy of Molecules on Surfaces, Methods of Surface Characterization*, Plenum Press, New York, 1987, pp. 345–412.
[44] B.E. Hayden, in J.T. Yates Jr and T.E. Madey (eds), *Vibrational Spectroscopy of Molecules on Surfaces, Methods of Surface Characterization*, Plenum Press, New York, 1987, pp. 267–340.
[45] S.A. Francis and A.H. Ellison, Infrared spectra of monolayers on metal mirrors, *J. Opt. Soc. Am.* 1959, **49**, 131–138.
[46] R.G. Greenler, Infrared study of adsorbed molecules on metal surfaces by reflection techniques, *J. Chem. Phys.* 1966, **44**, 310–315.
[47] M.K. Debe, Optical probes of organic thin films, *Prog. Surf. Sci.* 1987, **24**, 1–281.
[48] F.M. Mirabella Jr (ed.), *Internal Reflection Spectroscopy*, Marcel Dekker, New York, 1993.
[49] R. Aroca, S. Rodriguez-Llorente and M.D. Halls, Surface enhanced vibrational spectra of NTCDA on metal island films, *Asian J. Phys.* 1998, **7**, 391–404.
[50] J. Umemura, T. Kamata, T. Kawai and T. Takenaka, Quantitative evaluation of molecular orientation in thin Langmuir–Blodgett films by FT-IR transmission and reflection–absorption spectroscopy, *J. Phys. Chem.* 1990, **94**, 62–67.

3

Surface-Enhanced Raman Scattering

The interpretation of SERS spectra has, in many cases, been a frustrating experience for those simply attempting to obtain amplified Raman results quickly. This difficulty has a straightforward explanation: the observed SERS spectrum is a multivariate function of factors that in most cases are not possible to control, or worse, the experimenter is not aware of them. Hence it is important to examine and analyze closely the set of variables that may play a role in what is observed in SERS spectra.

Let us begin with the Raman vibrational spectrum of a molecule in gas phase. The basic components involved in this case are just the molecule and the incident radiation. To describe the observed Raman spectra we take the following steps: (i) the stationary vibrational energy levels of the molecule are determined; (ii) the incident radiation is defined in terms of its monochromaticity, polarization and intensity; and (iii) the dynamics of the interaction between the molecule and the incoming radiation field, the energy of the interaction and the selection rules that will establish the pattern of the observed spectrum are all determined. Notably, since there is a random orientation in the gas phase, directionality is not an issue in the spectral interpretation and polarization properties refer exclusively to the spatial orientation of the electric field of the light [1,2].

In contrast, in SERS, the basic components involved are a molecule, a metal nanostructure and electromagnetic radiation. This difference introduces a much greater degree of complexity to SERS experiments relative

Surface-Enhanced Vibrational Spectroscopy R. Aroca

to simple Raman measurements on gas samples. Let us document these important challenges to the interpretation of observed SERS spectra:

1. The molecule is interacting with a metal nanostructure. Using the well-established nomenclature of surface chemistry [3] and surface photochemistry [4], the adsorption on solid surfaces can be divided, according to the strength of bonding between the particle and the substrate, into two categories, physisorption and chemisorption. Physical adsorption (physisorption) refers to weak interactions arising from van der Waals forces, with adsorption energies well below those of normal chemical bonds. It is recognized that physical adsorption may alter the surface structure of molecular solids but not that of metals. When the adsorption energy is large enough and comparable to chemical bond energies (formation of a chemical bond), the term chemisorption is used.
2. Incident photons can induce substrate excitations such as electron–hole pairs, surface plasmons or surface phonons that may be involved in the enhancement of photo-induced processes. In particular, the absorption of light by nanostructures can create strong (enhanced) *local electric fields* at the location of adsorbed species, as discussed in Chapter 2. This new enhanced local field will strongly affect the optical properties of the adsorbate and is the main factor in bringing about the phenomenon of the giant SERS effect.
3. The interaction of incident radiation with adsorbed molecules may lead to photodissociation (and possible the creation of hot molecules), photoreactions or simply photodesorption. All of these processes can leave their own fingerprints in the observed SERS. The photodissociation of organic molecules on silver nanostructures was recognized early on, and is characterized by the well-known 'cathedral peaks' of SERS that arise from carbon products on silver.
4. The interaction of light with the metallic nanostructure depends on the value of the complex dielectric function at the excitation wavelength, and this will determine the enhancement observed at a given frequency of excitation. Since particle absorption and scattering (Chapter 2) depend on the shape and size of the metal nanostructure, SERS intensities are also influenced by these factors. In addition, the excitations in nanostructures are strongly influenced by the dielectric constant of the medium.
5. The dynamics of the interaction of light with the adsorbate leads to a pattern of Raman intensities determined by selection rules. Here

again, we should distinguish between the selection rules for vibrational transitions of a molecule in the gas phase, where infrared and Raman activities are easily determined from the symmetry point group, and the 'surface selection rules' for a 'fixed', spatially oriented molecule at the surface of an enhancing nanostructure. Surface selection rules encompass the symmetry properties of the dipole transitions and the modification of the intensities due to the components of the local electric field vector at the surface. They apply to molecules anchored on nanostructures where Raman and infrared intensities are further modulated by the spatial orientation of the local electric field (polarization) interacting with the polarizability derivative tensor (or dipole moment derivative). Since the adsorbed molecule generally belongs to a different symmetry point group than that of the parent molecule, the corresponding allowed modes and their polarization are also different. The interaction of polarized light with flat metal surfaces is predicted and calculated using Fresnel equations and coefficients (Chapter 2); however, the polarization of the local fields on metallic nanostructures is not that straightforward [5].

6. SERS is commonly obtained by excitation with visible or near-infrared light. The presence of the metal nanostructure may permit new excitations in the molecule–nanostructure complex, such as charge-transfer transitions from the Fermi to LUMO level of the molecule. Since the excitation is in resonance with the electronic transition of the adsorbed metal complex, the observed inelastic scattering is formally due to a related physical phenomenon: resonance Raman scattering. The observed SERS could indeed be surface-enhanced resonance Raman scattering (SERRS), and the observed relative intensities may not resemble the original Raman spectrum of the parent molecule.
7. Finally, small amounts of impurities may burst forth to give sudden signals that further complicate the interpretation of the observed SERS spectra [6].

The apparent controversial character of SERS is thus entirely due to the complexity brought about by the many factors that influence the observed spectrum. This complexity has led to a great deal of confusion in the literature and has hindered the development of quantitative analytical applications of SERS. In this respect, the lack of reliable and reproducible nanostructures with well-known enhancement factors has been, and continues to be, a major hurdle owing to the dispersion (frequency dependence) of enhancement. However, accumulated experience and extensive SERS databases are beginning to make further applications possible. It should be pointed out that many reports in the literature have significant flaws that must be understood and accounted for. On the positive side, despite the multiple variables associated with SERS, there are simple models that allow one to tackle experimental design and spectral interpretation. The main theory is developed around the electromagnetic enhancement mechanism that is discussed in the following section.

3.1 ELECTROMAGNETIC ENHANCEMENT MECHANISM

3.1.1 Definition of SERS

Kerker, in the introduction to his selected papers on SERS, comments on the link between the inelastic scattering of Raman and the elastic scattering of Mie: "The topography is complicated; the interconnections are manifold. Yet with the discovery of surface-enhanced Raman scattering we find two channels that link elastic scattering by metal colloids with inelastic scattering by molecules" [7]. A definition that is based on this link, describing the results explained with the simplest physical model, and capturing the essentials of the SERS phenomena, is due to Moskovits and co-workers [8]:

> As it is currently understood SERS is primarily a phenomenon associated with the enhancement of the electromagnetic field surrounding small metal (or other) objects optically excited near an intense and sharp (high Q), dipolar resonance such as a surface-plasmon polariton. The enhanced re-radiated dipolar fields excite the adsorbate, and, if the resulting molecular radiation remains at or near resonance with the enhancing object, the scattered radiation will again be enhanced (hence the most intense SERS is really frequency-shifted elastic scattering by the metal). Under appropriate circumstances the field enhancement will scale as E^4, where E is the local optical field.

The electromagnetic mechanism has been the fundamental theory for the computational approach to SERS enhancement and has been thoroughly reviewed in the literature. The most recent review by Schatz and Van Duyne [9] contains the pertinent references and is recommended.

Some aspects of the theory are being revisited in the literature and further development of specific topics, in particular that of 'hot spots', is ongoing [10]. Although there are treatments of the electromagnetic mechanism in SERS, from randomly rough metal surfaces using Gaussian statistics and correlation functions [11], from the very beginning surfaces used for SERS were viewed as nano-objects (surface protrusions) or isolated particles, as with colloids and metal island films. Because these model nano-objects are much smaller than the wavelength of the electromagnetic radiation used for excitation, the particles can be seen as being embedded in a static field and Maxwell's equations may be replaced by the LaPlace equation of electrostatics. Using the optical properties of the bulk materials for small nano-objects (another approximation), computational work on SERS properties is well on its way. However, the model systems used here to lay down the fundamentals for the interpretation of the observed SERS spectrum should be seen as they are, 'useful models', as much as the harmonic oscillator is an excellent model to help the understanding of complex molecular vibrations. The reality, brought to the forefront by experimental science, particularly in single molecule experiments or ultrasensitive detection, is most revealing of the strong connection between SERS and highly interacting metal nanoparticles, as has been demonstrated by observations in far-field and near-field experiments. Stretching our analogy of explaining the collective nature of molecular vibrations by the coupling of harmonic oscillators, the collective nature of the excitation in the SERS electromagnetic enhancement could also be approximated by the coupling of the individual plasmon resonances of the interacting nanoparticles.

3.1.2 Single Particle SERS Model Systems

The absorption and scattering of light by metallic nanoparticles (particles smaller than the wavelength of the light) is considered to be the most important property that gives rise to SERS. The excitation of the particle plasmon resonance, discussed in Chapter 2, provides the theoretical basis for the development of the electromagnetic enhancement mechanism [5,7,10,12–16]. We can go further to say that the plasmon resonance in nanostructures is a necessary condition for the observation of SERS. It defines the existence of SERS and, therefore, when this component is not present in the Raman scattering experiment, the results should not be classified as SERS. For instance, Raman scattering from flat, nonabsorbing surfaces should not be labelled SERS, and an effort should be made to regard these observations at the air–solid or liquid–solid interfaces formally as spontaneous Raman scattering or resonance Raman scattering.

The extinction and scattering by spheres are completely explained by exact Mie theory [17,18], with no need for approximations. However, an approximation to the Mie result can help with the understanding of the physics involved. Therefore, restricting the expansion of the scattering coefficients (Mie theory) to the first few terms, for metal particles smaller than the wavelength, the extinction and scattering are found to be proportional [Equation (2.29) in Chapter 2] to a g factor:

$$g = \frac{\varepsilon(\omega) - \varepsilon_m}{\varepsilon(\omega) + 2\varepsilon_m}$$

indicating that the computation is carried out with the dielectric function of the metal as a function of the frequency ω in a medium with dielectric constant ε_m, and that the g factor has a singularity that occurs at the plasmon resonance condition $\varepsilon(\omega) = -2\varepsilon_m$.

The same g factor appears in electrostatics when considering the problem of a small sphere of radius a embedded in a uniform static electric field, where the polarization of the sphere leads to an ideal dipole that is also given in terms of g: $p_0 = 4\pi\varepsilon_0 g a^3 E_0$. Therefore, an incident frequency ω_0 that fulfils the resonance condition $\varepsilon(\omega_0) = -2$ in vacuum, where $\varepsilon_m = 1$, leads to a very large factor

$$g_0 = \frac{\varepsilon(\omega_0) - 1}{\varepsilon(\omega_0) + 2}.$$

Since $\varepsilon(\omega)$ is complex, the resonance conditions is fulfilled for $\mathrm{Re}\{\varepsilon(\omega_0)\} = -2$, and the imaginary part will determine the quality factor of the resonance. For instance, the dielectric function of silver at 354 nm has $\mathrm{Re}\{\varepsilon(\omega)\} = -2.0$, giving rise to a resonance absorption centered at that wavelength.

Evidently, electromagnetic enhancement is independent of the presence of a molecule. In an early report on 'Photoinduced luminescence from the noble metals and its enhancement on roughened surfaces', Boyd *et al.* [19] clearly documented enhancement on rough films of Ag, Cu and Au. They pointed out that the "analysis of the effects of local field enhancement shows that the multiphoton luminescence is emitted predominantly from the surface atoms of protrusions on the rough surface with localized plasmon resonances at ω_1. The rise in the luminescence intensity towards lower energies is attributed to this resonance" [19].

3.1.3 Spherical Model

We consider single sphere–single molecule SERS at the surface of a spherical particle with dielectric function $\varepsilon(\omega)$ embedded in a medium of dielectric constant ε_{m}. The model for a molecule at a distance from the metal nanostructure and the model for chemical adsorption of the molecule onto the nanostructure are shown in Figure 3.1.

The discussion of the scattering of a spherical particle can be illustrated using the results obtained in electrostatics (particle in a static field) or results for particles small compared with the wavelength (electrostatic approximation), as in Chapter 2. The isolated sphere is illuminated and becomes polarized in the electromagnetic field owing to collective displacement of the electrons with respect to the nuclei, which is resonant at $\varepsilon(\omega) = -2\varepsilon_m$ (plasmon resonance). We can follow Kerker's model for the SERS [20–22] of a molecule placed outside the metal sphere as shown in Figure 3.1. The first step is to clarify and separate the different electric fields involved in the model. The incident oscillating field is a plane electromagnetic wave with electric field $E_0(r, \omega_0)$ at the point r and with frequency ω_0. According to Figure 3.1, the molecule is located in the *near field* of the sphere.

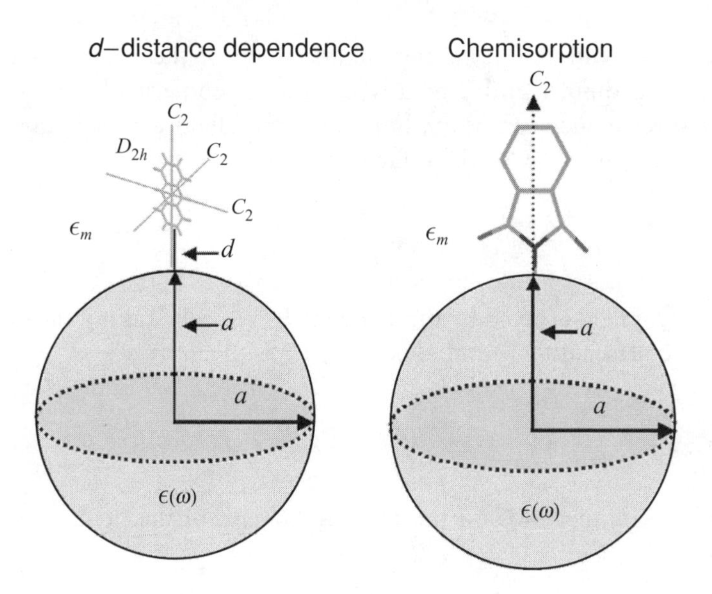

Figure 3.1 Cartoon representing a molecule at a distance d from the metal nanosphere and a molecule attached to the nanostructure or chemical adsorption model

3.1.3.1 *Polarization of the Sphere*

In the electrostatic problem of the sphere of radius a and dielectric function $\varepsilon(\omega)$, immersed in a dielectric with dielectric constant ε_m, to which uniform field is applied along the z-axis, where the center of the sphere is the origin of the coordinate system, the potentials Φ inside and outside the sphere are solutions of the Laplace equation [23]. Applying the boundary conditions to the general expression, the scalar potential outside the sphere is given by

$$\Phi_{\mathrm{OUT}} = -E_0 r\cos\theta + g\frac{a^3}{r^2}E_0\cos\theta. \tag{3.1}$$

The first term is potential due to incident field E_0, and is the result of the boundary condition, $(\Phi_{\mathrm{out}})_{r\to\infty} = -E_0 r\cos\theta$. The second term is equivalent to the potential due to an ideal dipole at the centre of the sphere, resulting from the polarization of the sphere with polarizability α. The induced dipole in the metal sphere in SI units is (reference 24, p. 158) $p_0 = \alpha E_0 = 4\pi\varepsilon_0 g a^3 E_0$, where the general polarizability is given by

$$\alpha = 4\pi\varepsilon_0\frac{\varepsilon(\omega) - \varepsilon_m}{\varepsilon(\omega) + 2\varepsilon_m}a^3 = 4\pi\varepsilon_0 g a^3.$$

The g factor will determine the enhancement of the polarizability and the induced dipole, and a greatly enhanced dipole field occurs at the plasmon resonance condition. For an external field with a frequency ω_0 and in vacuum ($\varepsilon_m = 1$), the g factor is simply

$$g_0 = \frac{\varepsilon(\omega_0) - 1}{\varepsilon(\omega_0) + 2}.$$

The local field is given by the gradient $E = -\nabla\Phi$. Taking the gradient in polar coordinates the radial component is given by

$$E_r = E_0\cos\theta a_r + 2g_0\frac{a^3}{r^3}E_0\cos\theta a_r. \tag{3.2}$$

The scattering depends on the absolute square of the field, and $E_r{}^2$ at $r = a$ is

$$E_r{}^2 = \left(1 + 4g_0{}^2 + 4g_0\right)E_0{}^2\cos^2\theta \quad \text{or} \quad E_r{}^2 = (1 + 2g_0)^2 E_0{}^2\cos^2\theta. \tag{3.3}$$

Since θ is the angle between the applied field direction and the vector r that locates positions on the sphere surface, it can be seen that the maximum enhancement is attained for $\theta = 0$ or 180. Also, the component $E_\theta{}^2 = 0$ at $\theta = 0$ or 180 and the local field is $E_r = (1 + g_0) E_0$. The scattering collected in the far field is proportional to $g_0{}^2$, and it is often named Rayleigh scattering for small particles, a parallel with molecular Rayleigh scattering. There are, however, some objections to associating Rayleigh's name with small particle scattering. A complete treatment of the scattering by a sphere with a historical postscript can be found in Chapter 3 of Kerker's book [18].

3.1.3.2 Polarization of the Molecule

For the isolated molecule, the Raman scattered field is due to the radiation of the induced electric dipole $p_M = \alpha_M E_0$, where α_M is the corresponding molecular polarizability. In spontaneous Raman, i.e. when Placzek's polarizability theory (2) can be used, the variation of the polarizability α_M with vibrations can be expressed by expanding each component of the tensor α in a Taylor series with respect to the normal coordinate: $\alpha_M = \alpha_0 + \alpha_0{}^1 Q_1 + \frac{1}{2}\alpha_0{}^{11} Q_1{}^2 + \ldots + \alpha_0{}^{12} Q_1 Q_2 + \ldots$, with

$$\alpha_0{}^1 = \left(\frac{\partial \alpha}{\partial Q_1}\right)_0 ; \alpha_0{}^{12} = \left(\frac{\partial^2 \alpha}{\partial Q_1 \partial Q_2}\right)_0$$

and where $Q_1 = Q_{10} \cos(\omega_1 t + \delta_1)$ is the normal coordinate. The first term is responsible for Rayleigh scattering, the second term, polarizability derivatives, gives the Raman scattering for fundamental vibrations and the other terms correspond to overtones and combinations.

3.1.3.3 Polarization of the Sphere–Molecule System

The observed Raman scattering from a molecule near a sphere will have two components. A molecule with polarizability $\alpha_{M,}$ located at the point of the local field E_r, can be described by a new induced oscillating dipole, $p_1 = \alpha_M E_r$. Note that the oscillation of p_1 is at the Raman frequency, ω. Notably, the polarizability and its derivatives can be strongly affected by the adjacent metal surface, altering the new induced dipole, and giving rise to several effects that are commonly grouped together and attributed to chemical enhancement effects. Since the numerical estimation of these contributions is elusive in most cases, exact enhancement factor values are case specific. These contributions will be discussed more completely in a separate section.

The second component to be observed in the far field is due to a secondary or scattered field. This field is associated with the new dipole p_1 and can be evaluated in exactly in the same way as E_r. Following Kerker, we call this field $E_{\mathrm{DIP}}(\omega)$, where the new frequency is the shifted Raman frequency of the molecule on the surface.

The new field $E_{\mathrm{DIP}}(\omega)$ polarizes the sphere and induces a dipole $p_2 = 4\pi\varepsilon_0 g a^3 E_{\mathrm{DIP}}$. The potential for this case is formally equivalent to that given by Equation (3.1):

$$\Phi_{\mathrm{out}} = -E_r r \cos\theta + g\frac{a^3}{r^2} E_r \cos\theta \tag{3.4}$$

and the new field E_{SC} is obtained by taking the gradient in polar coordinates:

$$E_{\mathrm{DIP}} = E_{\mathrm{r}} \cos\theta a_r + 2g\frac{a^3}{r^3} E_r \cos\theta a_r. \tag{3.5}$$

The square of the field is at $\theta = 0$:

$$E_{\mathrm{DIP}}{}^2 = \left(1 + 4g^2 + 4g\right) E_r{}^2 \quad \text{or} \quad E_{\mathrm{DIP}}{}^2 = (1 + 2g)^2 (1 + 2g_0)^2 E_0{}^2. \tag{3.6}$$

If we consider molecules located at the position of maximum enhancement ($\theta = 0$) under resonance conditions, the factor $(1 + 2g)(1 + 2g_0)$ captures the concept and gives a numerical approximation for the overall enhancement arising from the incident field E_0.

In the far field, or radiation zone, the detected Raman field for the sphere–molecule will have the contributions of the two oscillating dipoles p_1 and p_2. The complete evaluation of the fields was reported 25 years ago by Kerker *et al.* [22]. In summary, there are three oscillating induced dipoles: (1) p_0, due to the collective oscillation of the sphere with frequency ω_0 that can give enhanced Rayleigh scattering; (2) p_1 and (3) p_2, both induced dipoles oscillating with a shifted frequency ω, the inelastically scattered or Raman frequency. The p_1 dipole create its own new local dipolar field $E_{\mathrm{DIP}}(\omega_{\mathrm{Raman}})$ that is the first contributor to the observed Raman radiation. E_{DIP}, in turn induces a dipole p_2 in the sphere, also oscillating with the frequency ω (Raman frequency). The total Raman scattering observed in SERS, at a distance far from the sphere–molecule system, is due to the electric field created by p_1 [$E_{\mathrm{DIP}}(\omega_0)$] coherently added with the field [$E_{\mathrm{SC}}(\omega)$] given by p_2. The total power radiated is proportional to $|p_1 + p_2|^2$ and should be compared with the power radiated by the molecule in the absence of the sphere, $p_M{}^2$. Therefore, the enhancement factor is given by the ratio

$$EF = \frac{|p_1 + p_2|^2}{|p_M|^2}. \tag{3.7}$$

This enhancement factor does not include any changes in the value of the polarizability derivative (chemical effect) that can result from the molecule–metal proximity. Practical equations can be obtained from the square of the far-field amplitude to compute enhancement factors [16]. One of the expressions given by Kerker *et al.* (25) is

$$EF = 5\,|1 + 2g_0 + 2g + 4gg_0|^2\,. \tag{3.8}$$

$5[(1+2g)\,(1+2g_0)]^2 = 5(1+2g_0+2g+4gg_0)^2$ gives an example of a numerical evaluation of the SERS enhancement. One can see that the dominant term at the resonance frequency, the square of $4gg_0$ or $16g^2g_0{}^2$, is the approximation given in the review of Schatz and Van Duyne [9].

According to Equation (3.8), maximum SERS enhancement is achieved by excitation into the plasmon absorption of the nanoparticle used as a substrate. The result of a simple computation based on Equation (3.8) is given in Figure 3.2 for a silver sphere, showing an enhancement factor of 10^4. This figure illustrates the fact that, according to the model, *EF*

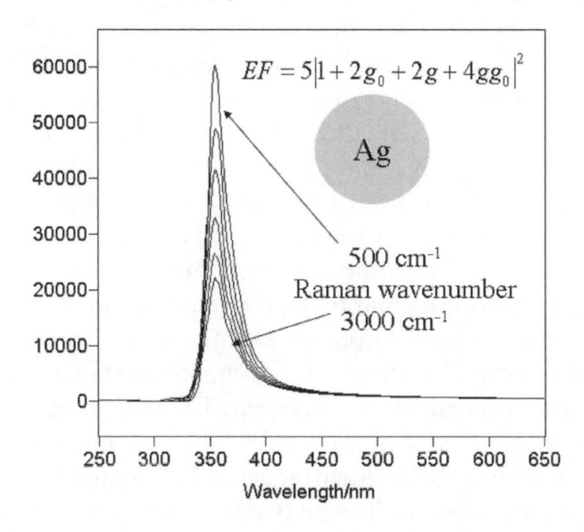

Figure 3.2 Enhancement factor calculations for a silver sphere showing an *EF* of 10^4, and illustrating the *EF* dependence on the vibrational wavenumber value

decreases with higher vibrational wavenumber values if the excitation is at the center of the plasmon resonance. Although in practice spheres may not be the best enhancers, they are the best pedagogical tool to illustrate the fundamentals of the physical phenomena. The resonance condition in the g factor, $\mathrm{Re}\{\varepsilon(\omega_0) + 2\} = 0$, clearly illustrates two of the most important properties of SERS:

1. SERS enhancement will be operative only for materials with appropriate optical constants in the spectral region of excitation (ω_0). Thereby, a quick browse of optical constants reveals that in the visible region the most commonly used coinage metals, silver, gold and copper, are good enhancers, and this is supported by the thousands of publications using these metals in the visible region of the spectrum to obtain SERS spectra.
2. The surface plasmon resonance condition for enhancing metal particles brings to the forefront the important role of the imaginary part of the $\varepsilon(\omega_0)$ in the magnitude of the g factor. Maximum enhancement is achieved if the imaginary part (damping) is small at the resonant frequency. For instance, the real and imaginary parts of ε (at 354 nm) for platinum are -3.9 and 4.1, respectively, indicating the lack of plasmon resonance in that region and a large damping value.

There is no reason to despair at the fact that only a few metals work. As we have seen in Chapter 2, the plasmon resonance can be modulated by the shape, size and external dielectric constant (ε_m). Therefore, the electromagnetic mechanism provides a guide to surface structure preparation for achieving SERS enhancement at particular excitation frequencies. This means that poor enhancers in the visible region may be able to provide adequate enhancement in the near-infrared region of the spectrum. SERS experiments are case specific and each problem requires tuning of the experimental conditions to maximize the signal-to-noise ratio or maximize the detection limit for a given analyte. Apart from the properties of the resonance condition, the equations for enhancement, based on this very simple model, also describe other properties and peculiarities of the electromagnetic SERS enhancement. They show that enhancement scales as the fourth power of the local field of the metal sphere, which is almost exactly the case for low-frequency Raman modes and explains why the scattering power of Raman bands fall off with increasing vibrational energy.

3.1.4 The Spheroidal Model

The electrostatic approximation (nano-objects much smaller than the wavelength of the incident radiation) has been used to calculate *EF* for a variety of models, including hemispheroids protruding from a conducting plane [26], isolated spheroids [see Gersten and Nitzan in the book on SERS edited by Chang and Furtak (reference 27, p. 89)] [28–31], and calculations for a two-sphere system [32] showing the strong dependence of the optical properties on the separation between the spheres. The electrostatic approximation assumes a uniform field inside the particle and is size independent. The introduction of corrections, for electrodynamic depolarization and damping effects, to the simple small particle Laplace electrostatic field gives results equivalent to solving Maxwell's equations, as done in a full electrodynamic solution [31]. In the work by Zeeman and Schatz [31], the dependence of both field and Raman enhancement factors on particle size and shape was studied.

Full electrodynamic computations can be carried out at a much greater computational cost. These are size-dependent calculations and include the multipole terms and the phase retardation from different sections of the object. The exact electrodynamic solutions for spheres was reported by Kerker *et al.* [25], including the small particle limits. Barber *et al.* [33] reported the first electrodynamic computation of *EF*s and electric field on the surface of Ag prolate spheroids showing that, as the size increases, the field enhancement decreases and the resonance shifts to the red, the resonance linewidth broadens and a new set of higher order resonances appears. They calculated the surface-averaged intensity and its wavelength dependence for a large $200 \times 100\ \text{nm}^2$ prolate spheroid embedded in the dielectric constants of air, water and cyclohexane. For small systems, the electrodynamic computations confirm the results obtained within the electrostatic approximation. In physical terms, this means that the dipolar term in the expansion dominates, i.e. the Rayleigh approximation is a good approximation for the complete Mie treatment of small objects. Therefore, the maximum field enhancement, $|E|_{\text{tip}}{}^2$, is found for small sizes (where the electrostatic solution is valid), and with size increases the enhancement decreases rapidly, the bandwidth broadens and the center of the resonance moves towards longer wavelengths. Notably, when the value of the dielectric constant of the medium increases, the enhancement also increases, and the resonance shifts further to the red.

The computational work [33] for isolated spheroids was in reasonably good agreement with the experimental results obtained for SERS of molecules adsorbed on Ag surfaces fabricated using microlithographic techniques consisting of arrays of isolated submicron Ag particles that were uniform in shape and size [34]. In many practical applications, the prolate and oblate model systems are used to discuss the SERS observed on metal island films, one of the three most commonly used types of SERS substrates in analytical applications (together with metal colloids and electrochemically prepared surfaces). Metal island films have a long history, and extensive research on the optical properties and theoretical modeling was done long before SERS was discovered. For example, in 1904, Maxwell-Garnett presented his work on 'Colours in metal glasses and in metallic films' (see reference 1 in Chapter 2) and in 1975 Weber and McCarthy [35] reported 'surface-plasmon resonance as a sensitive optical probe of metal-island films'. Many groups have used vacuum-evaporated metal island films [36–39], arrays of metal particles [40] or regular metallic structures generated by electron-beam lithography [41]. In our group, we have explored the use of vacuum-evaporated films of pure metals and mixed metals. Typical AFM images of In and Ag vacuum evaporated films are shown in Figure 3.3.

Simple electrostatic calculations using ellipsoidal shapes are useful for illustrating the properties of the EM effect in SERS. The results for the electrostatic problem of the field inside and outside an ellipsoidal body are relevant to SERS interpretation. As in the case of the sphere, the electric field of the incident radiation E_i produces a dipolar plasma oscillation. However, the local field outside the spheroid is no longer uniform, and the tip of a prolate spheroid will carry much higher field than the rest of the surface. Therefore, the enhancement at the tip is consistently

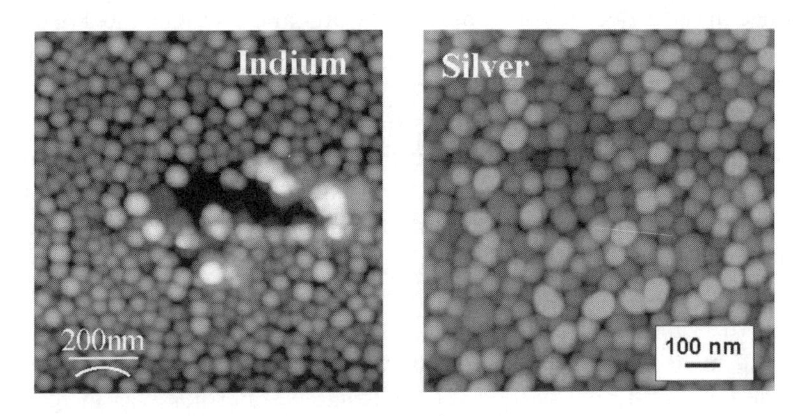

Figure 3.3 Atomic force microscopy images of indium and silver vacuum-evaporated films on glass commonly used in SERS experiments

higher than the average enhancement, and for very sharp tips the enhancement has been nicknamed the 'lightning-rod' effect (or the antenna effect). The treatment of the ellipsoid in the electrostatic approximation is given in Böttcher's book (reference 23, p. 79) and in Chapter 5 of Bohren and Huffman's book [17], where the potential inside (Φ_{ins}) and outside (Φ_{out}) the particle are given in terms of the external potential, the dielectric functions of the particle and the medium and the three depolarization (geometric) factors L [42] (A is also used in the literature for this geometric factor [28,30,33] and both L and A are used for the enhancement factor). In vacuum, the depolarization and geometric factors are identical. Most commonly, the model particle is a *prolate spheroid* (*cigar-shaped*), a spheroid where the axis of rotation is the major axis and the two equatorial geometrical factors are equal. Another common case is the *oblate*, where the axis of rotation is the minor axis. One can then consider fields applied parallel to the major or minor axes. For prolate spheroids, the plasmon resonances induced by the field applied along the minor axes occur at frequencies higher than that of the sphere, whereas resonances associated with fields along the major axis are at frequencies below that of the corresponding resonance of the sphere. It is assumed that it is the lower frequencies that contribute significantly to the observed SERS enhancements in spectroscopic measurements of metal island films containing prolate metal particles.

The treatment for a uniform field E_0, applied along the major axis (a), is similar to that described for a sphere. Formally, a dipole is induced, $p_L = \alpha E_0 = 4\pi\varepsilon_0 g^L ab^2 E_0$, given a potential outside the spheroid, where b is the minor axis and g^L represents a material factor, as we had for the sphere, modulated by L, the depolarization factor:

$$g_0{}^L = \frac{1}{3}\frac{\varepsilon(\omega_0) - \varepsilon_m}{\varepsilon_m + [\varepsilon(\omega_0) - \varepsilon_m]L}. \tag{3.9}$$

For a sphere with $L = \frac{1}{3}$, when $\varepsilon_m = 1$, g^L reduces to

$$g_0 = \frac{\varepsilon(\omega_0) - 1}{\varepsilon(\omega_0) + 2}.$$

Here again, we have two factors, one at the incident frequency and one at the scattered frequency ω:

$$g_s{}^L = \frac{1}{3}\frac{\varepsilon(\omega) - \varepsilon_m}{\varepsilon_m + [\varepsilon(\omega) - \varepsilon_m]L}. \tag{3.10}$$

When the external field is in the direction of the principal axis of the ellipsoid, then the field inside is

$$E_{x,y,z}^{\text{inside}} = \frac{1}{1 + \left[\varepsilon\left(\omega\right)/\varepsilon_m - 1\right] L_{x,y,z}} E_{x,y,z}^{\text{incident}}. \tag{3.11}$$

The behavior of a metal that is particularly relevant to SERS is the resonance conditions that enhance the field and the g factors. For a spheroidal metal particle, the resonance condition in these equations is achieved when the real part of the denominator vanishes:

$$1 + \left[\text{Re}\left\{\frac{\varepsilon\left(\omega_{\text{res}}\right)}{\varepsilon_m} - 1\right\}\right] L = 0. \tag{3.12}$$

The surface-average electromagnetic (EM) intensity enhancement factor is $A^2(\omega)$. Therefore, the overall enhancement detected in SERS measurements, according to this particular model, is approximately $A^2(\omega_0)$ $A^2(\omega_S)$ [27], where the subscript 0 refers to the frequency of the incident EM field and the subscript S refers to the Raman frequency. The total enhancement factor is given by the product of the g factors [28]:

$$\begin{aligned} A(\omega_0)A(\omega_s) &= 4g_0{}^L g_s{}^L \left(\frac{b^2}{a^2}\right)^2 = \frac{4}{3}\frac{\varepsilon(\omega_0) - \varepsilon_m}{\varepsilon_m + [\varepsilon(\omega_0) - \varepsilon_m]L} \\ &\quad \times \frac{1}{3}\frac{\varepsilon(\omega_s) - \varepsilon_m}{\varepsilon_m + [\varepsilon(\omega_s) - \varepsilon_m]L}\left(\frac{b^2}{a^2}\right)^2. \end{aligned} \tag{3.13}$$

The enhancement at the tip of the spheroid will give an additional multiplicative term, which is given by [28] γ^4, where $\gamma = \frac{3}{2}(a/b)^2\,(1 - L)$. Since L approaches zero as the aspect ratio increases, the value of γ increases rapidly with the aspect ratio. This is the enhancement due to the lightning-rod effect that already reaches $\gamma^4 = 2 \times 10^4$ for an aspect ratio of 3:1.

These two model systems can help facilitate the discussion of SERS and guide the experimental design for the most common substrates: metal colloids (spherical particles), metal island films (prolate and oblate) and rough electrode surfaces [43–46] (protrusion as semi-spheroids). In these model systems, molecules are treated as classical electric dipoles, electromagnetic fields are described by Maxwell's equations and metal particles are described by the optical properties of the bulk material.

Recently, the emphasis in the literature has shifted towards the fabrication of nanoparticles of different shapes such as triangles, nanowires, nanorods and aggregates of particles and fractals. Correspondingly, the

Figure 3.4 High-resolution transmission electron microscopy image of gold nanoparticles fabricated using chemical methods. It illustrates the possibility of controlling particle size and shape that can be achieved by varying the experimental conditions

models systems have been revisited and the field has become part of the general quest for the control and characterization of nanostructures as optical enhancers [47]. Monographs are appearing that reflect the pace of the growing field of nanostructure properties [48]. To illustrate nanofabrication using chemical methods, one image of many results obtained in our laboratory using fulvic acid in the synthesis of gold nanoparticles, where unprecedented control of particle size and shape is achieved by varying the experimental conditions, is shown in Figure 3.4 (high-resolution transmission electron microscopy of gold nanoparticles). Another very important implication is the fact that whereas spheres sustain one resonance and elliptical particles sustain only two resonances, triangular or particles of other shapes may display much more complex behavior with several resonances over a broad wavelength range.

3.1.5 The Shape Factor, Aggregates and Fractals

Spherical and spheroidal particles provide an excellent model system and exact solutions are available for their interactions with electromagnetic radiation (18). However, the experimental evidence accumulated over the years has pointed to the fact that the best enhancers are not spherical nanoparticles, but rather come in different shapes and particularly as aggregates of nanoparticles or fractal structures. Weitz *et al.* [49] in 1985 published their findings under the title 'Colloidal aggregation revisited: new insights based on fractal structure and surface-enhanced Raman scattering', pointing to the role of aggregation and fractal formation in the observed average SERS enhancement. Feilchenfeld and Siiman [50,51] also examined "a model in which a chain aggregate was constructed by sharing faces, edges, or corners of decahedra. Between each pair of adjacent particles in the chain there are wedge-shaped cavities of variable apical angle in which faces, edges and/or corners form parts of the wedge cavity. This particular set of shared faces, edges and corners may represent the SERS-active surface and also form a fractal curve. This is not inconceivable since the boundary of the face is multifaceted." A theory for SERS from fractals was proposed [52] and several approaches have been used that give an EM interpretation of SERS using fractal structures [53].

The EM interpretation is based on the idea of 'nano-resonators'. These nanostructures can be solid spheres, as in the model discussed above, solid structures of different shapes (ellipsoids, triangles, cylinders, etc.) or aggregates of nanoparticles or random fractals. It is immediately evident that whereas a sphere resonates at one particular frequency, a spheroid can resonate at three different frequencies, and a random fractal could resonate at any given wavelength within a broad spectral region [54,55]. Notably, the localized resonances in fractals carry large quality factors and seem to increase in number on proceeding from the visible to the near-infrared region. These elusive localized spots of small dimensions (smaller than the wavelength of the incident radiation) are the source of extremely highly enhanced local fields and correspondingly can produce SERS enhancement factors of 10^{12} or higher. These are the so-called 'hot spots' detected in SERS experiments.

In the original reports on single molecule detection, Nie and Emory [56] screened colloidal particles trying to pinpoint those that were optically hot (hot particles or 'hot spots'), whereas Kneipp *et al.* [57] probed individual silver clusters, and each of these silver clusters trapped a single dye molecule. It seemed that hot spots could be found on single particles as well as in clusters. This particular issue was examined experimentally by others groups [58,59] and the evidence was in favor of the aggregates as hot spots. Doering and Nie [60] revisited the problem of single particles versus aggregates and wrote that, "Recent results from both Kall's group and Brus's group indicate that the active sites for single-molecule SERS are likely located at the junction between two single particles", and added that, "it is possible that the halide-activated 'hot' particles observed in this study are also aggregates." The experimental evidence is accumulating and seems to indicate that the optically 'hot' particles are frequently aggregates consisting of two or more particles. There was another important difference in the 1997 reports on single molecule detection (SMD). One set of results was obtained using 514.5 nm laser radiation [56] to excite Rhodamine 6G (R6G), whereas Kneipp *et al.* [57] employed near-infrared excitation at 830 nm to excite the crystal violet molecule. The first excitation line is in resonance with the molecular electronic transition of R6G, whereas the second is far from it. The high enhancement obtained with the near-infrared excitation seems to contradict the idea of tuning into the surface plasmon of the metal particles. However, if we consider the aggregation and the near-infrared localized resonances found in fractals, large *EF*s in the near-infrared are expected. Although the exact correlation between 'hot spots' and surface plasmon is still under scrutiny, it is clear that for the observation of SERS an excitation in resonance with the Surface Plasmon absorption is necessary.

As has been already pointed out, the collective oscillation of the electrons in metal nanoparticles has been, and is, the object of intense multidisciplinary research with a wide range of applications. These oscillations are termed dipole particle plasmon resonances (DPPRs) (61) or simply surface plasmons (SPs). In particular, since the discovery of SERS, nanoparticles of silver and gold have been intensively investigated as the best known enhancers of optical signals. The role of the local field was seen for the case of the model systems. However, there is an important property of these fields that may lead to the explanations of the differences between the theoretical treatments and actual observed broad plasmon resonances, that is, the field produced by the plasmon resonances extends fairly far ($\sim$10 nm) from the surface. The implication of the interaction through these electromagnetic fields has been investigated since the initial formulation of the EM mechanism for SERS [32]. The computation of the very strong field variations that occur in plasmon resonant particles is a challenging problem and has been tackled by several groups. Kottmann *et al.* [62] reported plasmon resonances of non-regular silver particles in the 10–100 nm range and discussed their dependence on the particle shape and size, and also on the direction of illumination. Schatz's group [63] investigated the electromagnetic fields induced by optical excitation of localized surface plasmon resonances of silver nanoparticles and silver nanoparticle array structures that lead to exceptionally large electromagnetic field enhancements [64]. Stockman *et al.* [65] showed that surface plasmons of disordered nanosystems can have properties of both localized and delocalized states simultaneously. Etchegoin *et al.* [10] further examined the complex energy distribution and spatial localization of surface plasmon resonances in SERS active media, adding some physical insight to the idea of 'hot spots' by identifying them as regions with the largest concentration of electric field. They found that the red-shifted plasmon resonances in clusters of particles are responsible for the highest concentration of electromagnetic field lines. Theory and experiment are closing in on determining the strength and size of 'hot spots', and it is hoped will soon provide the understanding that will open the door to their reproducible fabrication.

3.1.6 Distance Dependence

Enhancement at a distance is a unique feature of EM enhancement that cannot be accounted for by any theoretical model based on the first layer effect. As illustrated in Figure 3.1, electromagnetic SERS enhancement does not require direct contact between molecule and metal, and our dipole description dictates that the field of a dipole decreases as $(1/d)^3$ with increasing distance. For the sphere, the decay of the enhancement to the fourth is given by $[a/(a+d)]^{12}$. This *distance dependence* has been measured using inert spacers layers in between the silver and the target molecule. Murray and Allara [29] used a polymer layer to separate the silver enhancing surface from *p*-nitrobenzoic acid (PNBA). They demonstrated the existence of a long-range enhancement above the rough silver films and concluded that: "Our results have implications for the use of surface-enhanced Raman scattering as a tool for measuring the vibrational spectrum of interfaces within ~100 Å of rough silver of the correct surface morphology". Wokaun and co-workers [14,66,67] reported a detailed experimental demonstration of the distance dependence of electromagnetic enhancement in surface-enhanced luminescence. As shown in Figure 3.5, taken from their work [67], the maximum fluorescence enhancement for basic fuchsin is attained at a well-defined thickness of the SiO_x spacer layer between the silver islands and the dye. Dots are experimental points and the full lines correspond to calculated values for silver spheroids of 30 × 14 nm (a) and silver prolate of 40 × 22.5 nm (b). Their experiments agree well with the theoretical model for EM enhancement as presented by Gersten and Nitzan [26], where SERS is maximum for the first layer, $d = 0$ in Figure 3.1, but fluorescence enhancement reaches its

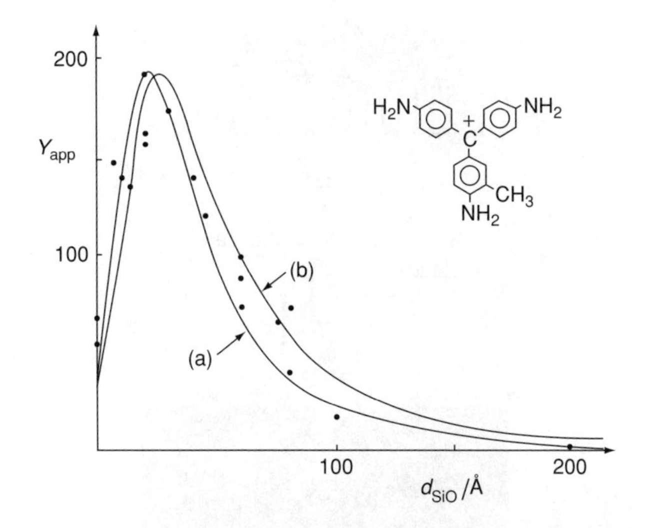

Figure 3.5 Experimental results showing that the maximum fluorescence enhancement for basic fuchsin is attained at a well-defined thickness provided by the SiO_x spacer layer between the silver islands and the dye. The full lines are calculated values using silver spheroidal model. Reproduced from A. Wokaun *et al.*, *Mol. Phys.* 1985, 56, 1–33 with permission from American Institute of Physics

maximum at a certain distance from the surface because at $d = 0$ energy transfers overpower the EM enhancement [68].

In our own group, we have carried out distance dependence studies for SERS [69] and surface-enhanced fluorescence (SEF) [70] using Langmuir–Blodgett (LB) monolayers of fatty acid as spacer layers, to separate the silver islands from the monolayer of the target molecule. All the experimental results for SERS and SEF agree with an enhancing effect detectable within ~100 Å of silver island films.

The LB technique in nanotechnology allows for the preparation of films with surface uniformity, nanometer thickness and molecular architecture manipulation. It is one of the most successful research techniques for the fabrication and study of organized molecular structures [71–73]. The conventional Langmuir technique normally employs typical amphiphilic molecules, i.e. molecules possessing distinct hydrophilic and hydrophobic parts that may interact strongly with water via hydrogen bonding, dipole–dipole interactions or dispersion forces. However, the technique has been extended to obtain uniform ultrathin films of nonconventional macromolecules such as polymers, various biomolecules, fullerenes and other macro heterocyclic compounds, for which the forces responsible for holding them on the water surface cannot be traced back to interactions with the water dipoles. Figure 3.6 shows typical

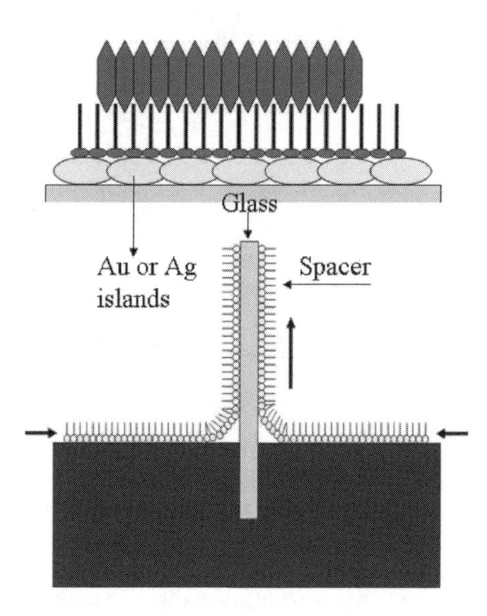

Figure 3.6 Cartoon representing Langmuir–Blodgett structures that can be built to study the distance dependence of EM enhancement in both SERS and SEF. A film of spheroidal nanoparticles fabricated on a glass substrate, a spacer monolayer of a fatty acid and the LB of target dye molecules.

structures that can be built to study the distance dependence of EM enhancement in both SERS and SEF. Silver and gold island films evaporated on to glass substrates are coated with LB structures of target molecules such as derivatives of the perylenetetracarboxylicdiimide (PTCD). These molecules provide the best example of enhancement of the Raman scattering and fluorescence in the same spectral range.

This is illustrated in Figure 3.7, where one LB monolayer deposited on a silver island film is excited with the laser radiation of 514.5 nm, a frequency in resonance with the molecular electronic transition of the PTCD molecule, thus giving rise to surface-enhanced resonance Raman scattering (SERRS)[74]. At the same time, the emission from the excimer fluorescence is just further down from the vibrational fundamentals, starting in the region of the first overtones and combinations. The LB monolayer on glass is the perfect reference (dashed line), since the spectra are obtained using Raman microscopy, and the surface area probed is the same in both spectra. The resonance Raman scattering (RRS) of the neat layer can not be seen above the background level (generally, in the author's experience, one can detect the RRS of a single PTCD monolayer fairly easily); however, the excimer fluorescence can be measured, as can be

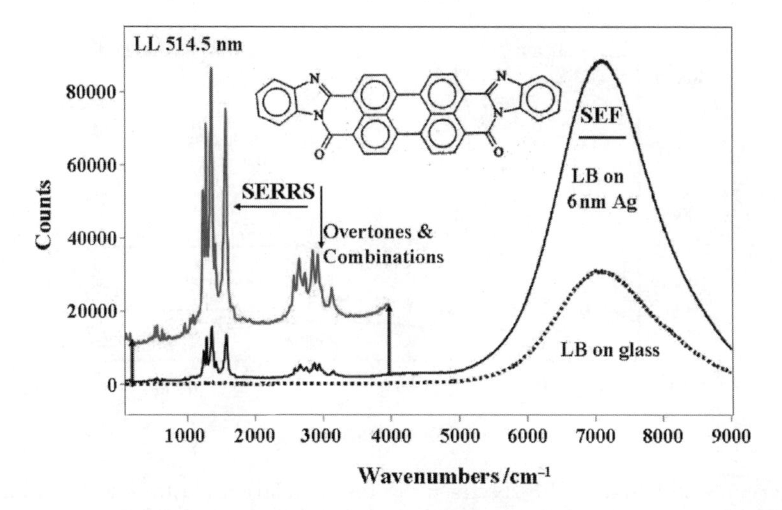

Figure 3.7 SERRS and surface-enhanced fluorescence in a single spectrum of an LB monolayer deposited on a silver island film when excited with the laser radiation of 514.5 nm. The excitation frequency is in resonance with the molecular electronic transition of the dye and with the surface plasmon absorption of the silver islands

seen in Figure 3.7. On silver islands the enhancement of both the Raman and excimer fluorescence are obvious.

With structures such as that shown in Figure 3.6, the distance dependence of Raman scattering and excimer fluorescence can be explored simultaneously in a single spectrum. The spacer monolayer film is fabricated using a fatty acid such as arachidic acid (AA) ($C_{19}H_{39}COOH$). These amphiphiles (one end is hydrophilic and the other is hydrophobic) form strong monolayers, and are consistently transferable with unit transfer ratios. The area per molecule occupied by AA on the surface is well known (25 Å^2), its hydrocarbon chain is chemically inert and it is a very poor scatterer in the region below the C–H stretching vibrations (below 2800 cm^{-1}). It provides an ideal matrix with a suitable window for vibrational studies using SERS/SERRS. The probe molecular system is a single LB monolayer of a PTCD derivative (monothio-BZP)[75] fabricated on glass, on silver and on top of one spacer layer of AA on silver. Laser radiation of 633 nm can be employed in the observation of SEF and SERRS from a monolayer, in the same spectrum, where the relative enhancement factors are directly comparable. The spectra for LB on glass (LB:glass) and LB on silver islands (LB:Ag) are also shown in Figure. 3.8. The average SEF enhancement factor was estimated from the two spectra to be about 10, where all the acquisition conditions have been kept constant. The SERRS enhancement obtained from the two spectra is ca 10^4.

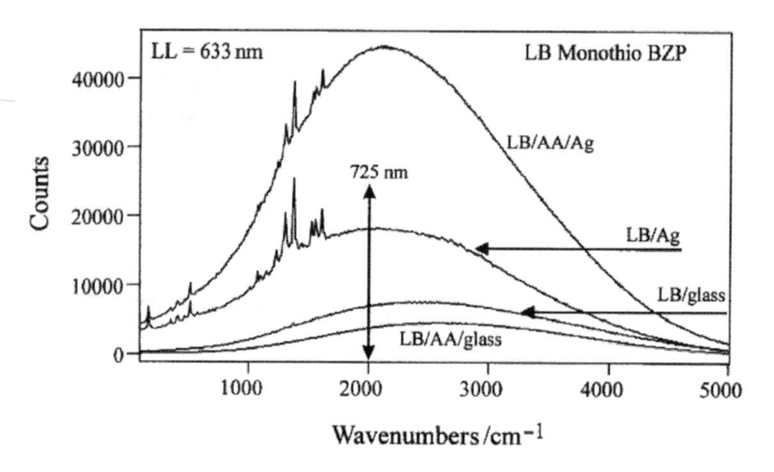

Figure 3.8 SERRS and SEF spectra using LB monolayers illustrating the distance dependence of the enhancement for scattering and fluorescence. Reproduced from C. Constantino, J. Duff and R. Aroca, *Spectrochim. Acta, Part A* 57, 1249–1259, 2001, with permission from Elsevier

These enhancement factors (*EF*) for the SEF and the SERRS of monothio-BZP are in agreement with the electromagnetic enhancement model for silver islands that makes it possible to compare the enhancement for Raman scattering (RS), resonance Raman scattering (RRS) and fluorescence, as described by Weitz *et al.* [76] In their work, a hierarchy of the 'average EM enhancement' was found for silver island films, with RS being about 10^5, RRS $\sim 10^3$ and the fluorescence enhancement depending on the quantum efficiency. The enhancement in SEF would be ~ 10 for molecules with low quantum efficiency.

3.1.7 Coverage Dependence of SERS

The interaction between resonances of particles separated from each other may help to bring new physical insights into the observations on aggregates and fractals. Similarly, the interactions between the point dipoles used to describe the spectroscopic properties of the adsorbed molecules will have an effect on the plasmon resonances and the observed intensities in the SERS effect. The point dipole moment is a crude approximation since it cannot account for the interactions between the polarized molecule and the electrons in the metal. The latter is the subject of a separate chapter on chemical effects (Chapter 4). The electromagnetic treatment of the enhancing nanoparticle that supports collective

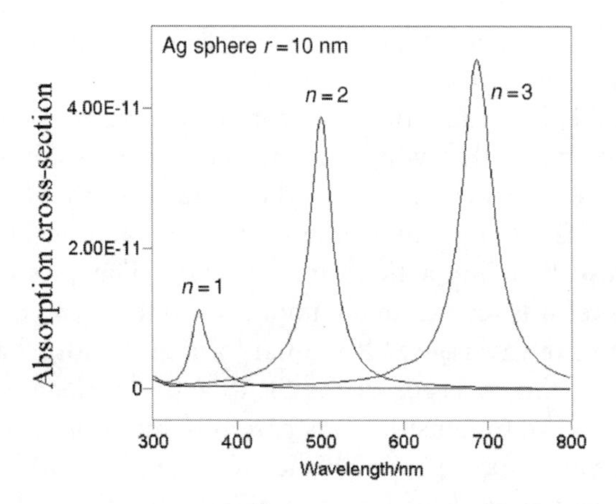

Figure 3.9 Calculations illustrating the effect on the plasmon absorption for a single silver sphere of 10 nm radius embedded in a solvent with variable refractive index from 1 (vacuum) to 3

electron oscillations immediately allows the prediction of at least two consequences of coating the particle with a dielectric (a coat of dipoles). First, the plasmon absorption will be affected by the dielectric coat with a dielectric constant different from 1, and the center of the absorption will be shifted. Let us start with two extreme cases. First, a single molecule on the particle will not have any significance effect on the optical properties of the metal particle. The opposite should be true for the optical properties of the molecule. The other extreme is a single particle embedded in a solvent with refractive index n, and the effect on the plasmon absorption is illustrated in Figure 3.9 for a silver sphere of 10 nm radius, embedded in medium with refractive index variable from 1 (vacuum) to 3. The dielectric constant of the medium changes significantly both the magnitude and the location of the plasmon absorption, strongly affecting the SERS signal due to the EM mechanism.

The general problem with a coated sphere or ellipsoid, where the thickness of the coating is variable, is discussed in Bohren and Huffman [17]. However, the ideal model should predict the effect on the SERS signal when molecules (point dipoles) are added to the surface of the enhancing nanoparticle up to the formation of a monolayer coverage. In chemical terms, it would be important to determine the variation of the SERS signal with the fraction of the surface covered, which for any given adsorbate, at any given temperature, is the *adsorption isotherm* [77]. The adsorption isotherm summarizes the adsorption properties of any given molecule on the SERS-active substrate. For instance, for an adsorption isotherm of the Langmuir type, characterized by a monotonic approach to the limiting adsorption of a complete monolayer, the observed SERS intensity would also follow the same pattern. However, in the Langmuir model, it is assumed that while the molecules occupy sites on the surface, they do not interact with each other. The observed SERS intensities do not follow the Langmuir isotherm pattern, hinting at an effective adsorbate–adsorbate interactions at the surface. The question of the adsorption isotherm is an extremely important one when considering the optimization of the average SERS signal for a given adsorbate.

The effect of the coverage of the SERS-active surface with the analyte on the observed intensities was carefully monitored from the very beginning of SERS. Otto [78] published a review in 1980, where he discussed (p. 324) the work of Rowe *et al.* [79] of SERS: "from pyridine absorbed on Ag surfaces in ultrahigh vacuum show a strong dependence on both *surface roughness and pyridine coverage*." The authors separated a SERS signal for the first layer from that of the physisorbed pyridine, and observed an enhancement (10^4) for physisorbed pyridine multilayers in addition to the first layer. They concluded that the enhanced effect was electromagnetic rather than chemical in origin. Similar experiments by Pockrand and Otto [78] found that the signal for the physisorbed species was about 10^2, but they proposed that atomic-scale roughness provides unique adsorption sites responsible for most of the enhancement. The first review on the subject of surface coverage by Sanda *et al.* can be found in the book on SERS edited by Chang and Furtak (reference 27, p. 189). The SERS experiments discussed were conducted using samples prepared in ultrahigh vacuum (UHV), and they had the advantage of working with *clean* substrates. The verdict from this work was that both the directly attached adsorbates and the long-range enhancement contribute to the observed SERS. In 1980, Seki and Philpott [80], working with pyridine and silver island films in UHV, reported that only the first layer contributes to the SERS signal, suggesting that certain special sites 'active sites' would be responsible for the entire SERS effect. The search for these sites or atomic-scale roughness and adatoms generated a great deal of SERS literature and Otto *et al.* [15] provided an excellent review of this work.

The treatment of the surface coverage according to the classical electromagnetic model for SERS was presented by Chew *et al.* [81] and the theoretical model revealed the dipole–dipole interaction of molecules adsorbed on the surface of a sphere. The numerical results presented for Ag particles of radius 10, 50 and 100 nm showed that the Raman signal decrease by about 15% as the dipoles approach each other on the surface of the Ag sphere. The coverage dependence of the average SERS signal was also studied independently by two groups [82,83], using the LB technique that allows one to control the coverage and to work with mixed monolayers containing well-defined concentrations of the probe molecule. The conclusion from both groups was the same: there is an optimum surface coverage for which the SERS is maximized and it corresponds, for single monolayer coverage, to a mixed monolayer with submonolayer coverage of the probe molecule. For example, the coverage dependence of SERS and SERRS studied using mixed LB monolayers of tetra-*tert*-butylvanadylphthalocyanine [(*t*-Bu)4VOPc] and arachidic acid on silver-coated Sn spheres give a maximum SERRS for a ca 25% molar ratio of the probe in arachidic acid. In the same work, the long-range effect was probed. "LB experiments with two consecutive LB layers (containing 20% arachidic acid) showed an increase in the SERS and SERRS signals. The contribution of the second layer is clearly due to the long-range nature of the electromagnetic enhancement". However, further LB coverage decreased the SERS and SERRS intensities. Since all SERS and SERRS signals observed were mainly due to electromagnetic enhancement, variation in intensity may be attributed to the fact that the dye layer strongly damps the resonance of the metal particle. Experiments with multilayers of arachidic acid:dye = 10:1 molar ratio showed a continuous increase in SERRS intensity with the number of mixed layers coating the Ag surface. The effect tapers off after the fourth layer, in agreement with a long-range enhancement due to surface plasmons. The mix LB layer technique emerged as a very powerful analytical tool for the study of strongly absorbing dyes using SERRS. For practical applications, the most efficient SERRS signal may be obtained by working with two or three mixed LB layers of dye (in the 10:1 to 5:1 region of molar ratio) on a SERS-active surface.

The adsorption coverage may lead to different orientations of the adsorbate on the surface and, correspondingly, spectral differences in the relative intensities are observed. Alimardonov *et al.* [84] looked at the changes in the SERS spectra of C_2H_6 adsorbed on cold deposited (85 K) silver film with the surface coverage, temperature and laser radiation time. They reported three adsorption forms of C_2H_6, formed step-by-step with increasing surface coverage. However, they may be identified with the physisorbed dissociation paths. An ethane–ethylene transition is found. This transition is photostimulated by laser radiation when the surface coverage is small, and it seems to be catalytic when the surface coverage is large. The results highlight the difficulties ever present in the interpretation of SERS spectra collected in the far-field, which include all the information about the different phenomena taking place on the surface.

Electrochemical methods, such as rapid linear sweep voltammetry, provide a way to measure the surface concentration of electroactive metal complexes, permitting one to determine the cross-section for average SERS and normal Raman scattering, and the corresponding enhancement factor [85]. Weaver *et al.* [85] found, for average SERS at silver electrodes, an *EF* of ca 10^6 for species such as NCS^-, $Cr(NCS)_6{}^{3-}$, pyridine, pyrazine and $Co(NH_3)_6{}^{3+}$. In this particular set of experiments, the authors reached the conclusion that "... surface attachment ('chemisorption') of the Raman scatterer to the metal may in itself exert little influence upon the enhancement factor." This particular issue will be discussed further in Chapter 4 on chemical effects, however, it should be pointed out that chemisorption may be desirable in order to obtain the maximum benefit from the local field enhancement (first-layer effect). The caveat of the surface coverage is that partial coverage of the surface would be more effective than full coverage owing to damping by dipole–dipole interactions.

The surface coverage was also probed by effectively covering the SERS-active Ag surface with a SERS-'inactive' metal such as Pb. Guy *et al.* [86] studied the effects of submonolayer and monolayer amounts of Pb deposited at underpotential on an Ag electrode surface on the SERS spectra of pyridine and Cl^-. It was found that surfaces covered by >70% of a Pb monolayer do not show surface enhancement. Lopez-Rios and Gao [87] carried out similar work using palladium.

Brolo *et al.* [88] investigated the inelastic scattering from pyrazine, adsorbed on smooth and rough gold electrodes. The variables were the surface coverage and surface morphology. The SERS intensity of the pyrazine ring-breathing mode (ca 1016 cm^{-1}) was followed as a function of the surface morphology (roughness factor). The SERS intensity was maximized between 20 and 30 oxidation–reduction cycles (ORCs). AFM measurements indicated that roughness features with an average size of ca 100 nm were present in the surface yielding the strongest SERS signal. The dependence of the SERS intensity on the surface morphology agrees well with the electromagnetic calculations of the enhancement factor for gold spheres. The Raman intensity from pyrazine adsorbed on a flat gold electrode follows the surface concentration up to about two-thirds of a monolayer. As the amount of pyrazine in the surface approaches the complete monolayer, this dependence is no longer observed, probably owing to molecular interactions on a heavily packed surface.

Within certain levels of concentration, the Langmuir adsorption isotherm has been found to be a good model for SERS intensity versus surface coverage. Loo *et al.* [89] discussed SERS spectra of benzotriazole (BTA) and 6-tolyltriazole (6-TTA) mixtures on Cu electrodes. There findings indicated that fractional coverage of these molecules on Cu surfaces is directly correlated with their absolute concentration in solution and is consistent with an assumed Langmuir adsorption isotherm model. The adsorption equilibrium constant of 6-TTA is about three times that of BTA, and the free energy of adsorption for 6-TTA on Cu is 610 cal mol^{-1} lower than that for BTA.

In summary, spectroscopy based on the inelastic scattering from a system of molecules adsorbed on nanoparticles comes with a long list of potential factors that will have an effect on the observed SERS intensity to the extent that reproducibility and quantitative analysis remain experimentally challenging. The observed SERS spectra can be classified into two groups, 'average' SERS obtained from an ensemble enhancing system and single molecule–single nanoparticle SERS. For all the SERS observations, the electromagnetic mechanism as the origin of and main contribution to the observed SERS intensity provides guidelines for experiment and spectral interpretation. The most important components of the EM models include:

- The excitation of a plasmon resonance in an isolated nanoparticle, void, aggregate of nanoparticles or surface roughness.
- The plasmon resonances are determined by the optical properties of the material, the shape of the nanostructure or roughness and other variables as in bimetallic structures, nanoshells or nanowires.
- The local field enhancement is maximum for a molecule directly attached to the surface (first-layer effect) and it also has a long-range component that extends its effect up to about 10 nm away from the interface.
- The role of multiple plasmon resonances is essential for the treatment of aggregates or closely spaced nanostructures.
- The surface coverage experiments and dipole–dipole interactions model show a deviation of the observed SERS intensity from the simple Langmuir adsorption isotherm. In fact, reproducibility and quantitative analysis for average SERS require a more detailed knowledge of the adsorption isotherms, as will be shown in Chapter 6.

REFERENCES

[1] E.B. Wilson Jr, J.C. Decius and P.C. Cross, *Molecular Vibrations; the theory of Infrared and Raman Vibrational Spectra*, McGraw-Hill, New York, 1955.
[2] D.A. Long, *The Raman Effect*, John Wiley & Sons, Ltd, Chichester, 2001.
[3] A.W. Adamsom, *Physical Chemistry of Surfaces*, John Wiley & Sons, Inc., New York, 1990.

[4] R. Franchy, Surface and bulk photochemistry of solids, *Repo. Progr. Phys.* 1998, **61**, 691–753.
[5] M. Moskovits, *Rev. Mod. Phys.* 1985, **57**, 783–826.
[6] S. Sanchez-Cortes and J.V. Garcia-Ramos, Anomalous Raman bands appearing in SERS, *J. Raman Spectrosc.* 1998, **29**, 365–371.
[7] M. Kerker (ed.), *Selected Papers on Surface-enhanced Raman Scattering*, SPIE, Bellingham, WA, 1990.
[8] C. Douketis, T.L. Haslett, Z. Wang, M. Moskovits and S. Iannotta, Self-affine silver films and surface-enhanced Raman scattering: linking spectroscopy to morphology, *J. Chem. Phys.* 2000, **113**, 11315–11323.
[9] G.C. Schatz and R.P. Van Duyne, in J.M.Chalmers and P.R. Griffiths (eds), *Handbook of Vibrational Spectroscopy*, John Wiley & Sons, Ltd, Chichester, 2002, pp. 759–774.
[10] P. Etchegoin, L.F. Cohen, H. Hartigan, R.J.C. Brown, M.J.T. Milton and J.C. Gallop, Electromagnetic contribution to surface enhanced Raman scattering revisited, *J. Chem. Phys.* 2003, **119**, 5281–5289.
[11] J.A. Sanchez-Gil, J.V. Garcia-Ramos and E.R. Mendez, Electromagnetic mechanism in surface-enhanced Raman scattering from Gaussian-correlated randomly rough metal substrates, *Opt. Express* [online computer file], 2002, **10**, 879–886.
[12] H. Metiu, T.E. Furtak and R.K. Chang (eds), *Surface Enhanced Raman Scattering*, Plenum Press, New York, 1981, pp. 1–34.
[13] M. Kerker, Electromagnetic model for surface-enhanced Raman scattering (SERS) on metal colloids, *Acc. Chem. Res.* 1984, **17**, 271–277.
[14] A. Wokaun, H.-P. Lutz, A.P. King, U.P. Wild and R.R. Ernst, Surface enhancement of optical fields. Mechanism and applications, *Mol. Phys.* 1985, **56**, 1–33.
[15] A. Otto, I. Mrozek, H. Grabhorn and W. Akemann, Surface-enhanced Raman scattering, *J. Phys. Condens. Matter* 1992, **4**, 1143–1212.
[16] A. Otto, Surface-enhanced Raman scattering of adsorbates, *J. Raman Spectrosc.* 1991, **22**, 743–752.
[17] C.F. Bohren and D.R. Huffman, *Absorption and Scattering of Light by Small Particles*, John Wiley & Sons, Inc., New York, 1983.
[18] M. Kerker, *The Scattering of Light and Other Electromagnetic Radiation*, Academic Press, New York, 1969.
[19] G.T. Boyd, Z.H. Yu and Y.R. Shen, Photoinduced luminescence from the noble metals and its enhancement on roughened surfaces, *Phys. Rev. B* 1986, **33**, 7923–7936.
[20] D.S. Wang, H. Chew and M. Kerker, Enhanced Raman scattering at the surface (SERS) of a spherical particle, *Appl. Opt.* 1980, **19**, 2256–2257.
[21] M. Kerker, Estimation of surface-enhanced Raman scattering from surface-averaged electromagnetic intensities, *J. Colloid Interface Sci.* 1987, **118**, 417–421.
[22] M. Kerker, D.S. Wang and H. Chew, Surface-enahced Raman scattering (SERS) by molecules adsorbed at spherical particles: errata, *Appl. Opt.* 1980, **19**, 4159–4173.
[23] C.J.F. Böttcher, *Theory of Electric Polarization*, Elsevier, Amsterdam, 1973.
[24] D.J. Jackson, *Classical Electrodynamics*, John Wiley & Sons, Inc., New York, 1999.
[25] M. Kerker, D.S. Wang and H. Chew, Surface-enhanced Raman scattering (SERS) by molecules adsorbed at spherical particles: errata, *Appl. Opt.* 1980, **19**, 4159–4173.
[26] J. Gersten and A. Nitzan, Electromagnetic theory of enhanced Raman scattering by molecules adsorbed on rough surfaces, *J. Chem. Phys.* 1980, **73**, 3023–3037.
[27] R.K. Chang and T.E. Furtak (eds), *Surface Enhanced Raman Scattering*, Plenum Press, New York, 1982.

[28] P.F. Liao and A. Wokaun, Lightning rod effect in surface enhanced Raman scattering, *J. Chem. Phys.* 1982, **76**, 751–752.

[29] C.A. Murray and D.L. Allara, Measurements of the molecule–silver separation dependence of surface-enhanced Raman scattering in multilayered structure *J. Chem. Phys.* 1982, **76**, 1290–1303.

[30] C.A. Murray, Simple model for estimating the electromagnetic enhancement in surface-enhanced Raman scattering from molecules adsorbed onto metal particles coated with overlayers, *J. Opt. Soc. Am. B* 1985, **2**, 1330–1339.

[31] E.J. Zeman and G.C. Schatz, An accurate electromagnetic theory study of surface enhancement factors for silver, gold, copper, lithium, sodium, aluminum, gallium, indium, zinc, and cadmium, *J. Phys. Chem.* 1987, **91**, 634–643.

[32] P.K. Aravind, A. Nitzan and H. Metiu, The interaction between electromagnetic resonances and its role in spectroscopic studies of molecules adsorbed on colloidal particles or metal spheres, *Surf. Sci.* 1981, **110**, 189–204.

[33] P.W. Barber, R.K. Chang and H. Massoudi, Electrodynamic calculations of the surface-enhanced electric intensities on large silver spheroids, *Phys. Rev. B* 1983, **27**, 7251–7261.

[34] P.F. Liao, Surface enhanced Raman scattering from lithographically prepared microstructures, *Springer Ser. Opt. Sci.* 1981, **30**, 420–424.

[35] W.H. Weber and S.L. McCarthy, Surface-plasmon resonance as a sensitive optical probe of metal-island films, *Phys. Rev. B* 1975, **12**, 5643–5650.

[36] T.M. Cotton, J.H. Kim and G.D. Chumanov, Application of surface-enhanced Raman spectroscopy to biological systems, *J. Raman Spectrosc.* 1991, **22**, 729–742.

[37] R.P. Van Duyne, J.C. Hulteen and D.A. Treichel, Atomic force microscopy and surface-enhanced Raman spectroscopy. I. Silver island films and silver film over polymer nanosphere surfaces supported on glass, *J. Chem. Phys.* 1993, **99**, 2101–2115.

[38] C. Jennings, R. Aroca, A.M. Hor and R.O. Loutfy, Surface-enhanced Raman scattering from copper and zinc phthalocyanine complexes by silver and indium island films, *Anal. Chem.* 1984, **56**, 2033–2035.

[39] Y.M. Jung, H. Sato, T. Ikeda, H. Tashiro and Y. Ozaki, Surface-enhanced Raman scattering study of organic thin films of 2-octadecyl-7,7,8,8-tetracyanoquinodimethane and its charge transfer films doped with 3,3′,5,5′-tetramethylbenzidine, *Sur. Sci.* 1999, **427–428**, 111–114.

[40] T.R. Jensen, M.D. Malinsky, C.L. Haynes and R.P. van Duyne, Nanosphere lithography: tunable localized surface plasmon resonance spectra of silver nanoparticles, *J. Phys. Chem. B* 2000, **104**, 10549–10556.

[41] M. Kahl, E. Voges, S. Kostrewa, C. Viets and W. Hill, Periodically structured metallic substrates for SERS, *Sens. Actuators B* 1998, **51**, 285–291.

[42] R. Aroca and F. Martin, Tuning metal island films for maximum surface-enhanced Raman scattering, *J. Raman Spectrosc.* 1985, **16**, 156–162.

[43] A.G. Brolo, D.E. Irish and B.D. Smith, Applications of surface enhanced Raman scattering to the study of metal-adsorbate interactions, *J. Mol. Struct.* 1997, **405**, 29–44.

[44] Z.-Q. Tian and B. Ren, Adsorption and reaction at electrochemical interfaces as probed by surface-enhanced Raman spectroscopy, *Annu. Rev. of Phys. Chem.* 2004, **55**, 197–229.

[45] J.E. Pemberton, A.L. Guy, R.L. Sobocinski, D.D. Tuschel and N.A. Cross, Surface enhanced Raman scattering in electrochemical systems: the complex roles of surface roughness, *Appl. Surf. Sci.* 1988, **32**, 33–56.

[46] R.L. Garrell, Surface-enhanced Raman spectroscopy, *Anal. Chem.* 1989, **61**, 401A–402A, 404A, 406A–408A, 410A–411A.

[47] S. Link and M.A. El-Sayed, Shape and size dependence of radiative, non-radiative and photothermal properties of gold nanocrystals, *Int. Rev. Phys. Chem.* 2000, **19**, 409–453.

[48] D.L. Feldheim and C.A. Foss (eds), *Metal Nanoparticles. Synthesis, Characterization and Applications*, Marcel Dekker, New York, 2002.

[49] D.A. Weitz, M.Y. Lin and C.J. Sandroff, Colloidal aggregation revisited: new insights based on fractal structure and surface-enhanced Raman scattering, *Surf. Sci.* 1985, **158**, 147–164.

[50] H. Feilchenfeld and O. Siiman, Adsorption and aggregation kinetics and its fractal description for chromate, molybdate, and tungstate ions on colloidal silver from surface Raman spectra, *J. Phys. Chem.* 1986, **90**, 4590–4599.

[51] O. Siiman and H. Feilchenfeld, Internal fractal structure of aggregates of silver particles and its consequences on surface-enhanced Raman scattering intensities, *J. Phys. Chem.* 1988, **92**, 453–464.

[52] M.I. Stockman, V.M. Shalaev, M. Moskovits, R. Botet and T.F. George, Enhanced Raman scattering by fractal clusters: scale-invariant theory, *Phys. Rev. B* 2002, **46**, 2821–2830.

[53] J.A. Sanchez-Gil and J.V. Garcia-Ramos, Strong surface field enhancements in the scattering of p-polarized light from fractal metal surfaces, *Opt. Communi.* 1997, **134**, 11–15.

[54] V.A. Shubin, A.K. Sarychev, J.P. Clerc and V.M. Shalaev, Local electric and magnetic fields in semicontinuous metal films: beyond the quasistatic approximation, *Phys. Rev. B* 2000, **62**, 11230–11244.

[55] V.M. Shalaev, Surface-enhanced optical phenomena in nanostructured fractal materials, *Handb. Nanostruct. Mater. Nanotechnol.* 2000, **4**, 393–449.

[56] S. Nie and S.R. Emory, Probing single molecules and single nanoparticles by surface-enhanced Raman scattering, *Science* 1997, **275**, 1102–1106.

[57] K. Kneipp, Y. Wang, H. Kneipp, L.T. Perelman, I. Itzkan, R.R. Dasari and M.S. Feld, Single molecule detection using surface-enhanced Raman scattering (SERS), *Phys. Rev. Lett.* 1997, **78**, 1667–1670.

[58] A.M. Michaels, M. Nirmal and L.E. Brus, Surface-enhanced Raman spectroscopy of individual Rhodamine 6G molecules on large Ag nanocrystals, *J. Am. Chem. Soc.* 1999, **121**, 9932–9939.

[59] H. Xu, E.J. Bjerneld, M. Kall and L. Borjesson, Spectroscopy of single hemoglobin molecules by surface enhanced Raman scattering, *Phys. Rev. Lett.* 1999, **83**, 4357–4360.

[60] W.E. Doering and S. Nie, Single-molecule and single-nanoparticle SERS: examining the roles of surface active sites and chemical enhancement, *J. Phys. Chem. B* 2002, **106**, 311–317.

[61] K.L. Kelly, E. Coronado, L.L. Zhao and G.C. Schatz, The optical properties of metal nanoparticles: the influence of size, shape, and dielectric environment, *J. Phys. Chem. B* 2003, **107**, 668–677.

[62] J.P. Kottmann, O.J.F. Martin, D.R. Smith and S. Schultz, Field polarization and polarization charge distributions in plasmon resonant nanoparticles, *New J. Phys.* 2000, **2**, 27.1–27.9.

[63] E. Hao and G.C. Schatz, Electromagnetic fields around silver nanoparticles and dimers, *J. Chem. Phys.* 2004, **120**, 357–366.

[64] S. Zou and G.C. Schatz, Silver nanoparticle array structures that produce giant enhancements in electromagnetic fields, *Chem. Phys. Lett.* 2005, **403**, 62–67.

[65] M.I. Stockman, S.V. Faleev and D.J. Bergman, Localization versus delocalization of surface plasmons in nanosystems: can one state have both characteristics?, *Phys. Rev. Lett.* 2001, **87**, 167401/1–167401/4.

[66] A. Wokaun, H.P. Lutz and A.P. King, Distance dependence of surface enhanced luminescence, *Springer Ser. Chem. Phys.* 1983, **33**, 86–89.

[67] A. Wokaun, H.P. Lutz, A.P. King, U.P. Wild and R.R. Ernst, Energy transfer in surface enhanced luminescence, *J. Chem. Phys.* 1983, **79**, 509–514.

[68] C.D. Geddes and J.R. Lakowicz (eds), Topics in Fluorescence Spectroscopy. Vol. 8, *Radiative Decay Engineering*, Springer, New York, 2004.

[69] G.J. Kovacs, R.O. Loutfy, P.S. Vincett, C. Jennings and R. Aroca, Distance dependence of SERS enhancement factor from Langmuir–Blodgett monolayers on metal island films: evidence for the electromagnetic mechanism, *Langmuir* 1986, **2**, 689–694.

[70] R. Aroca, G.J. Kovacs, C.A. Jennings, R.O. Loutfy and P.S. Vincett, Fluorescence enhancement from Langmuir–Blodgett monolayers on silver island films, *Langmuir* 1988, **4**, 518–521.

[71] A. Ulman, *Ultrathin Organic Films*, Academic Press, Boston, 1991.

[72] G.L. Gaines, *Insoluble Monolayers Liquid–Gas Interfaces*, Interscience, New York, 1966.

[73] G. Roberts, *Langmuir–Blodgett Films*, Plenum Press, New York, 1990.

[74] C.J.L. Constantino and R.F. Aroca, Surface-enhanced resonance Raman scattering imaging of Langmuir–Blodgett monolayers of bis(benzimidazo)perylene on silver island films, *J. Raman Spectrosc.* 2000, **31**, 887–890.

[75] C. Constantino, J. Duff and R. Aroca, Surface enhanced resonance Raman scattering imaging of Langmuir–Blodgett monolayers of bis(benzimidazo)thioperylene, *Spectrochim. Acta, Part A* 2001, **57**, 1249–1259.

[76] D.A. Weitz, S. Garoff, J.I. Gersten and A. Nitzan, A comparison of Raman scattering, resonance Raman scattering, and fluorescence from molecules adsorbed on silver island films, *J. Electron Spectrosc. Related Phenom.* 1983, **29**, 363–370.

[77] D.M. Ruthven, *Principles of Adsorption and Adsorption Processes*, John Wiley & Sons, Inc., New York, 1984.

[78] A. Otto, Surface enhanced Raman scattering (SERS), what do we know?, *Appl. Surf. Sci.* 1980, **6**, 309–355.

[79] J.E. Rowe, C.V. Shank, D.A. Zwemer and C.A. Murray, Ultrahigh-vacuum studies of enhanced Raman scattering from pyridine on silver surfaces *Phys. Rev. Lett.* 1980, **44**, 1770–1773.

[80] H. Seki and M.R. Philpott, Surface-enhanced Raman scattering by pyridine on silver island films in an ultrahigh vacuum, *J. Chem. Phys.* 1980, **73**, 5376–5379.

[81] H. Chew, D.S. Wang and M. Kerker, Effect of surface coverage in surface-enhanced Raman scattering: Interaction of two dipoles, *Phys. Rev. B* 1983, **28**, 4169–4178.

[82] R. Aroca and D. Battisti, SERS of Langmuir–Blodgett monolayers: coverage dependence, *Langmuir* 1990, **6**, 250–254.
[83] J.H. Kim, T.M. Cotton, R.A. Uphaus and D. Moebius, Surface-enhanced resonance Raman scattering from Langmuir–Blodgett monolayers: surface coverage–intensity relationships, *J. Phys. Chem.* 1989, **93**, 3713–3720.
[84] E. Alimardonov, A.N. Gass, O.I. Kapusta and S.A. Klimin, Surface-enhanced Raman spectra of ethane adsorbed on silver film. Three adsorption forms of ethane. The ethane–ethylene transition, *Poverkhnost* 1987, 10–20 (in Russian).
[85] M.J. Weaver, M.A. Farquharson and M.A. Tadayyoni, Surface enhancement factors for Raman scattering at silver elctrodes. Role of adsorbate–surface interactions and electronic structure, *J. Chem. Phys.* 1985, **82**, 4867–4874.
[86] A.L. Guy, B. Bergami and J.E. Pemberton, The effect of underpotentially deposited lead on the surface enhanced Raman scattering ability of polycrystalline silver. *Surf. Sci.* 1985, **150**, 226–244.
[87] T. Lopez-Rios and Y. Gao, Modification by palladium adsorbates of the surface enhanced Raman scattering at silver surfaces, *Surf. Sci.* 1988, **205**, 569–590.
[88] A.G. Brolo, D.E. Irish, G. Szymanski and J. Lipkowski, Relationship between SERS intensity and both surface coverage and morphology for pyrazine adsorbed on a polycrystalline gold electrode, *Langmuir* 1998, **14**, 517–527.
[89] B.H. Loo, A. Ibrahim and M.T. Emerson, Analysis of surface coverage of benzotriazole and 6-tolyltriazole mixtures on copper electrodes from surface-enhanced Raman spectra, *Chem. Phys. Lett.* 1998, **287**, 449–454.

4

Chemical Effects and the SERS Spectrum

4.1 PHYSICAL AND CHEMICAL ADSORPTION

After the local electromagnetic field enhancement (EM mechanism) of the Raman scattering intensity has been discussed, the effect of the electronic interactions between the metal and the adsorbate should be considered in order to tackle the interpretation of the observed Raman spectra. Recently, Otto wrote [1], "For sure, without the EM mechanism there would be no signal. But the chemical mechanism determines what is observed." This comment captures the fundamental fact that the observed SERS spectrum contains the information about the adsorbate and its environment, in particular its interaction with the enhancing nanoparticle, its spatial orientation and the polarization properties of the local electric field.

In this chapter, we examine the Raman vibrational spectrum of *admolecules*, molecules adsorbed on an active SERS nanostructure. The factors affecting the rate of adsorption are not discussed here (see Chapter 6), nor are the effects induced by the photon flux (photodesorption and photoreactions), so we examine the Raman spectrum of an adsorbate that does not degenerate during illumination. The electronic cloud of the adsorbate represented by the volume $\alpha(Q, \omega)$ can be distorted by the interaction with the solid surface of the nanostructure. When the energy of this interaction (enthalpy of adsorption) is more positive than ~ -25 kJ mol^{-1}, the interaction is classified as physisorption, and the

Surface-Enhanced Vibrational Spectroscopy R. Aroca

physisorption potential has a d^{-3} dependence with the molecule–surface distance d. When the enthalpy of adsorption is very negative and comparable to chemical bond energies (formation of a chemical bond), we use the term chemisorption [2,3]. There is a gray area for the distinction between physisorption and chemisorption; however, when the interaction energy is more negative than $-40\,\text{kJ mol}^{-1}$, it is assumed that chemisorption has taken place. Chemisorption imposes a dramatic change in the volume $\alpha(Q,\omega)$ of small molecules. In fact, the molecular property $\alpha(Q,\omega)$ becomes that of a new molecule or surface complex, with different point group symmetry (different vibrational spectrum) and new electronic states.

Physisorption also distorts the volume $\alpha(Q,\omega)$, and correspondingly the first derivatives of $\alpha(Q,\omega)$ could be different from those of the isolated molecule. Since the first derivatives of $\alpha(Q,\omega)$ determine the Raman intensities, the pattern of relative intensities within the observed SERS spectrum for physisorbed species may differ from that of the reference molecule. In addition to the changes in polarizability derivatives, adsorption may induce a fixed molecular orientation. Experimentally, the spontaneous Raman scattering (RS), the total Stokes scattered light, averaged over all random molecular orientations I_{RS} (photons s^{-1}), is directly proportional to the incoming flux of photons I_0 (photons s^{-1} cm^{-2}), $I_{\text{RS}} = \sigma_{\text{RS}} I_0$ (the proportionality constant σ_{RS} is the Raman cross-section). A fixed molecular orientation introduces *directionality*, meaning that with the use of polarized light, the symmetry species of the allowed Raman modes may be distinguished. The situation is similar to that of the molecular vibrations in crystals [4]. However, given the fact that the metal surface strongly polarizes the light, the local field can be seen as the sum of a component perpendicular to the surface (p-polarized) and a component parallel or tangential to the surface (s-polarized). Factoring in the latter variables after the normal irreducible representation has been done results in the 'surface selection rules' discussed in Chapter 3.

The SERS spectra of chemisorbed species require a brand new vibrational analysis to be started, since one may be facing a completely different vibrational spectrum from that of the neat adsorbate molecule. Thereby, one will start by assuming a given geometry for the chemisorbed surface complex, finding its symmetry point group, carrying out the irreducible representation analysis and finding the species of symmetry. Finally, the molecular orientation of the surface complex and the light polarization at the surface of the nanostructure will permit one to carry out the interpretation of the observed SERS spectrum. In summary, one is facing a very different vibrational spectrum in terms of both frequency

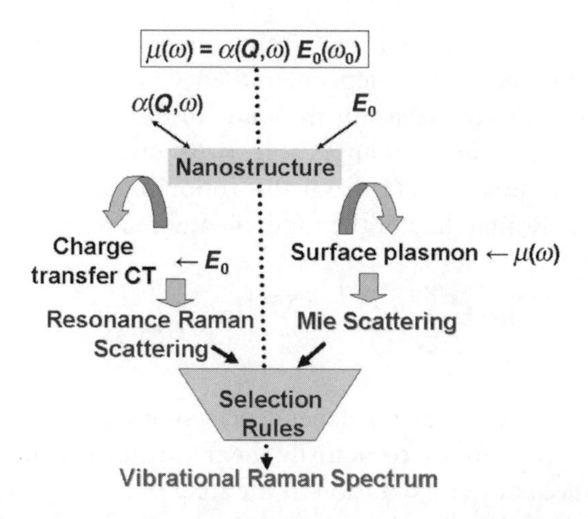

Figure 4.1 Flow chart showing the most significant of different contributions that can influence the observed SERS spectrum. Emphasis is placed on the key sieving role of the selection rules that tune the observed relative Raman intensities to a few normal modes

values and relative intensities. However, from the point of view of the enhancement, the direct attachment to the surface provides the highest EM enhancement factor giving rise to 'the first-layer effect' [5]. This new SERS spectrum can be compounded by the possibility of a multiplicative enhancement generated by a resonance Raman effect due to excitation into a charge-transfer [6] (CT) transition, as was pointed out at the very beginning of the discovery of SERS [7,8].

A flow chart for the step-by-step characterization is presented in Figure 4.1. The classical induced dipole is at the top. On the right-hand side the cartoon represents the EM contribution. On the left, the chemical effects are represented as changes in the $\alpha(Q,\omega)$; especially in the first derivative of its Taylor's series expansion that determines the Raman intensities. It should be remembered that the effects are multiplicative. In this chapter, we intend to examine and illustrate with examples each of the most commonly encountered cases: SERS of physically adsorbed molecules, SERS of chemically adsorbed molecules without CT resonance, chemically adsorbed molecules with CT resonance and surface-enhanced resonance Raman scattering (SERRS).

The special case of *resonance Raman scattering* (RRS), in the diagram, occurs whenever the sample is irradiated with exciting radiation whose energy corresponds to that of the electronic transition in the molecule, a chromophore within a large molecule or a CT transition. Under this resonance condition, the intensities of some Raman bands originating in this chromophore may be selectively enhanced by several orders of magnitude. The frequency-independent theory is Placzek's polarizability theory that provides the basis for the study of Raman scattering (RS) by molecules. Placzek's theory is applicable to excitations in the transparent region of the substance, i.e. far from absorption bands. In these cases, the intensity of the Raman line of the fundamental modes (I_{SC}), given by [9]

$$I_{SC} = \frac{8\pi}{9c^4} (\omega_{SC})^4 \sum_{m,n} \left|(\alpha_{\rho,\sigma})_{m,n}\right|^2 I_0 \tag{4.1}$$

shows that the only frequency dependence comes from the factor ${\omega_{SC}}^4$. Equation (4.1) describes satisfactorily the frequency dependence for colorless substances on RS excitation in the long-wavelength region of the visible spectrum. Notably, deviations from the proportionality between the Raman line intensity and the ${\omega_{SC}}^4$ value are also found for colorless materials [10], an indication that pre-resonant contributions to the Raman amplitude from higher electronic states are often significant. However, a dramatic deviation is only seen when the frequency dependence of the polarizability is activated. In this case, the polarizability is determined by the sum of the contributions of the electronic excited levels. The frequency-dependent polarizability $(\alpha_{\rho,\sigma})_{m,n}$ is a tensor component for the m to n transition, with incident and scattered polarization indicated by ρ and σ, respectively. An expression for the tensor component can be obtained using second-order perturbation theory [11]:

$$(\alpha_{\rho_\sigma})_{m,n} = \frac{1}{h}\left(\sum_e \frac{\langle m|\, D_\sigma \,|e\rangle \langle e|\, D_\rho \,|n\rangle}{\nu_{em} - \nu_L - i\Gamma_e} + \frac{\langle m|\, D_\rho \,|e\rangle \langle e|\, D_\sigma \,|n\rangle}{\nu_{en} + \nu_L - i\Gamma_e}\right) \tag{4.2}$$

Where D is the electronic dipole moment operator, $|e\rangle$ corresponds to the intermediate state, $i\Gamma_e$ is the damping term giving the homogeneous width of the intermediated state and ν_L is the laser frequency. After a Herzberg–Teller expansion of the electronic wavefunctions, the expressions are grouped into the Albrecht's terms: $(\alpha_{\rho\sigma})_{m,n} = A + B + C$. Qualitatively, the vibrational modes that are vibronically active in the electronic transition should be seen with high intensity in the RRS spectrum. Albrecht's A term is comprised of vibrational overlap factors and the squares of these factors are called Franck–Condon factors; the corresponding Raman term is Frank–Condon scattering. When the shape of the potential and equilibrium geometry of the excited state and ground state are different, the Frank–Condon A term may determine the Raman scattering and overtones progressions are observed [12]. There are many study cases in the literature for diatomics, where the effect of the *A* term in the RRS spectrum has been clearly illustrated [12]. In polyatomics, the *A* term makes a minor contribution and the Herzberg–Teller *B* term dominates the scattering. Nevertheless, Frank–Condon scattering with its corresponding overtone progressions can be observed in polyatomic molecules. This is illustrated in Figure 4.2 with the RRS spectrum of the lapis lazuli, a mineral from Chile, and that of bis(benzylimido)perylene (Bbip-PTCD) excited at 244 nm. In both cases the overtone progressions are clearly observed. Since the ground state is generally totally symmetric, only totally symmetric vibrations act as intermediates, and they become enhanced in the RRS spectrum of the *A* term case. However, the Bbip-PTCD derivative has a second electronic absorption seen in the visible region, in resonance with the 514.5 nm laser line. In this case, the *B* term may become important (Herzberg–Teller interaction), and the intermediate state no longer has to be totally symmetric. Therefore, totally and nontotally symmetric vibrations may be enhanced in the RRS spectrum. This is illustrated in Fig. 4.2 by the RRS spectrum of Bbip-PTCD

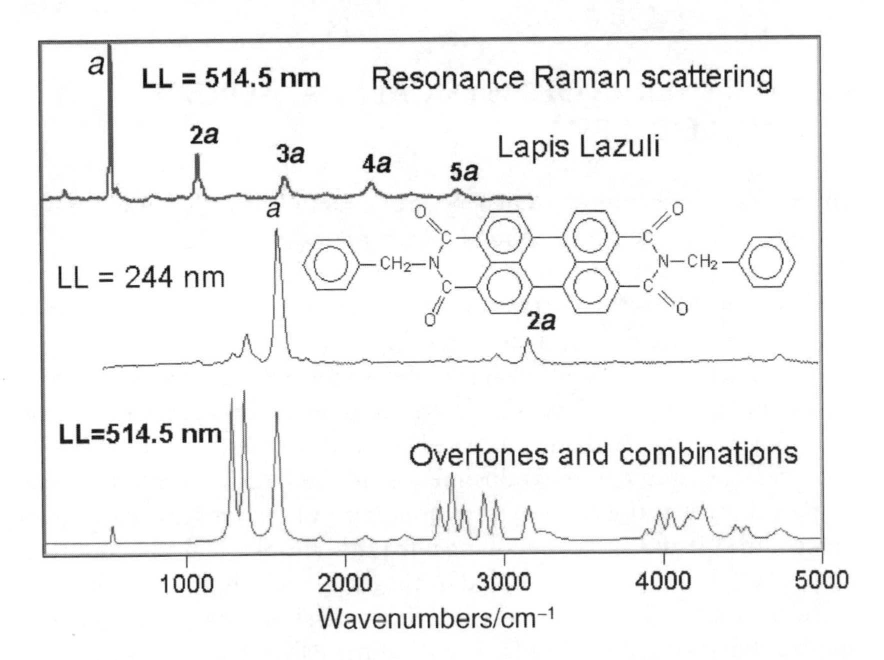

Figure 4.2 Illustration of the resonance Raman effect showing fundamental vibrational modes and overtone and combinations for an inorganic material (top) and excitation of two electronic resonances within the same organic material (bottom)

excited in resonance with an electronic transition in the visible region with the 514.5 nm laser line. It can be seen that nontotally symmetric ring vibrations are also present with the corresponding combinations and overtones progressions.

Interpreting the RRS spectrum of a polyatomic is in itself a challenging proposition. In the review of the time-dependent resonant Raman theory [13], Myers warns, "...there are a number of other reasons why resonance Raman intensity analyses can go astray even if the experimental data are very good." SERRS inherits all of these challenges. However, with the problems also comes the good news of enhanced Raman cross-sections in the RRS experiments. For instance, for dyes absorbing in the visible region, the RRS intensity can be 10^4–10^6 times that of the non-resonant spontaneous RS. It can be seen that, in principle, by selecting to work with an excitation frequency that produces the RRS effect in the probe molecule and surface-enhanced effect on an active metal nanostructure, the multiplicative effect can easily provide enhancements in the intensity of the order of 10^{10}. That could be sufficient for single molecule detection [14].

4.2 SERS/SERRS OF PHYSICALLY ADSORBED MOLECULES

This is perhaps the only case when we can retain the expectation of obtaining a SERS spectrum that resembles that of the reference, the normal Raman or resonant Raman spectra. As an example of physisorption we used here the Bbip-PTCD molecule, which can be manipulated to form a single monomolecular layer using the LB technique. A single LB monolayer was deposited on a glass slide to record the RRS spectrum shown at the top in Figure 4.3. There is a strong fluorescence background and the spectrum has been baseline corrected in order to see the Raman band. The overtones and combinations are lost in the fluorescence. The same LB procedure was used to extend a monolayer of the probe on to a silver island film [14] and an indium island film [15]. The physical adsorption of Bbip-PTCD is indirectly confirmed by the metal quenching of the fluorescence and the SERRS spectra revealed the strong intensity of the overtone and combinations of the fundamentals of the PTCA chromophore.

It can be seen in Figure 4.3 that the SERRS spectrum on silver islands is an enhanced version of the RRS spectrum excited with 514.5 nm laser radiation. The minor changes in the relative intensity of the fundamentals

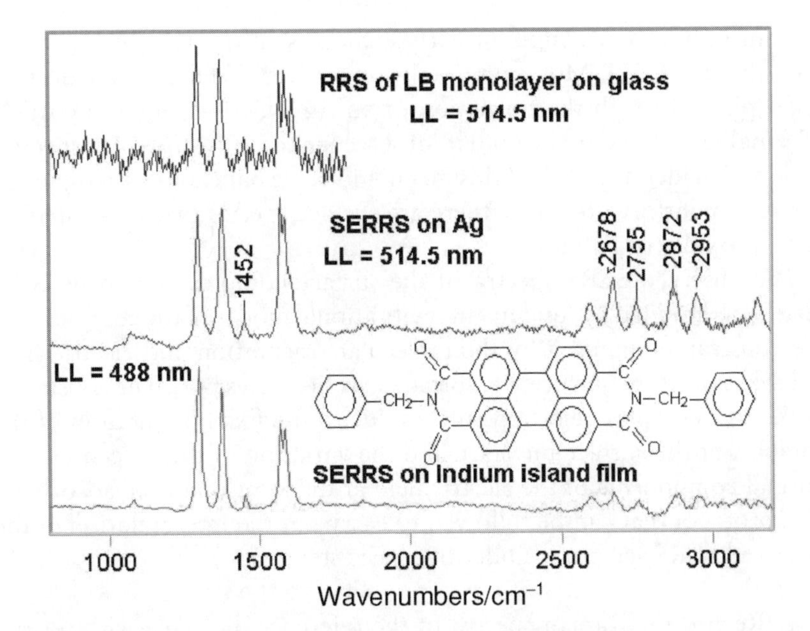

Figure 4.3 Resonance Raman scattering (top) and enhanced-resonance Raman scattering of a physisorbed organic molecule on silver and indium island films. The enhanced spectra are from a single LB monolayer coating the metal islands

can be explained by the fact that the sample is an LB monolayer, where there is a degree of molecular organization. The intensity of overtones and combinations is maximized by excitation at 514.5 nm, a result that was obtained by studying the dispersion of the chromophore with laser lines from 325 to 785 nm. Therefore, the relative intensities observed on indium island films at 488 nm correspond with what is observed on Ag, and this is not an effect of the metal.

4.3 SERS OF CHEMICALLY ADSORBED MOLECULES WITHOUT ELECTRONIC RESONANCE EXCITATION

Here we examine the interpretation of the SERS spectrum of a chemically adsorbed molecule that forms a *surface complex*, but it does not have an electronic absorption in the spectral region of the excitation, i.e. there is no resonance Raman effect. In the case of chemisorption, identification of the bonding and characterization of the surface complexes in SERS are the fundamental tasks in order to interpret the observed spectra, and we can only give a couple of early examples of the vast literature on the subject [16,17]. Most commonly, on a silver surface, chemisorption takes place through the formation of an Ag–N, Ag–S, Ag–O or Ag–X (X = halogen) bond that would be observed in the 100–300 cm^{-1} spectral region. The identification of this metal–molecule band and its assignment is not straightforward, since there are surface species that may interfere such as oxides or halides.

The observed SERS spectra of the surface complex should be completely determined by fundamental vibrational modes of the complex and the constraints imposed by the molecular orientation and electric field polarization at the surface, i.e. surface selection rules [18]. The constraint in the free complex selection rules is due to the fixed orientation of the complex at the surface interface and the variation of the tangential and normal components of the electric field at the location of the adsorbate.

A protocol that can be followed to carry out the interpretation of the observed SERS spectra includes five basic steps:

1. Record the Raman spectra of the reference molecule with several laser lines (excitation spectrum) to rule out contributions from the resonant Raman effect. Select one laser line for the experiments.
2. Synthesize the salt that corresponds to the possible surface complex and record the Raman spectra with the same laser line as in 1. Measure its UV–visible absorption spectrum and the excitation spectrum.
3. Record the SERS spectra with the same laser line as in 1.
4. Compute theoretically the Raman intensities for the reference molecule and the complex.
5. Apply the constraints of the surface selection rules to computed spectra for direct comparison with the observed SERS.

The use of quantum chemical calculations to model SERS spectra is today a powerful and common analytical technique [19–22]. Interpretation of the experimental data is facilitated by progress in the efficiency of programs for computing derivative properties, such as normal-mode frequencies and spectral intensities, and advances in computer hardware that have made quantum chemical calculations an invaluable aid to spectroscopists in the analysis and assignment of experimental vibrational spectra for complex polyatomics. Density functional theory (DFT) in recent years has proved to be superior to conventional *ab initio* methods in computing quantitative vibrational properties in a cost-effective fashion. Work by Halls *et al.* [23] has shown that DFT predicts harmonic frequencies in excellent agreement with observed fundamentals, having empirical scaling factors close to unity (see also Scott and Radom [24]). Studies of the vibrational intensities furnished by DFT methods lend additional support to the application of DFT in computing reliable theoretical vibrational spectra to assist in the interpretation of spectra for unique and complex molecular systems. Thereby, the introduction of simple computational models and their validation using archetypal chemisorbed analytes is a powerful analytical tool for spectral interpretation. According to Halls *et al.* [23], scaling factors determined from regression analysis for obtaining fundamental vibrational frequencies calculated at six of the most commonly used levels of theory show that, overall, the best performers are the hybrid functional theory (DFT) methods, Becke's three-parameter exchange functional with the Lee–Yang–Parr fit for the correlation functional (B3-LYP) and Becke's three-parameter exchange functional with Perdew and Wang's gradient-correlation functional (B3-PW91). For Raman intensities, the polarized-valence triple-zeta (pVTZ) Sadlej electronic property basis set is particularly useful. Hybrid DFT methods along with the Sadlej pVTZ basis set provide reliable theoretical vibrational spectra in a cost-effective manner.

To illustrate the case of a chemical adsorption without apparent resonant Raman effect, let us consider the spectra of thiophenol, C_6H_6S, an aromatic molecule with $N = 13$ and 33 normal modes of vibration. The characteristic high-frequency vibrations of the neat material are the S–H stretching and the C–H stretches, above 2500 cm^{-1}. These normal modes happen to carry a high relative intensity, as can be seen in Figure 4.4.

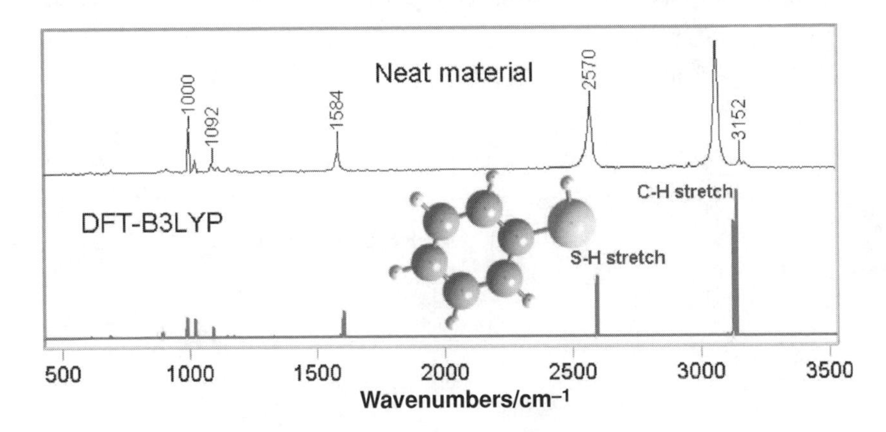

Figure 4.4 Experimental and calculated Raman spectra of thiophenol (C_6H_6S) illustrating the good agreement attained for an aromatic molecule with 33 normal modes of vibration

The assignment is readily helped by the *ab initio* calculated Raman intensities and frequencies. It can be seen that there is reasonable agreement between the calculated and the experimental spectra. Considering that the neat material is not in the gas phase, the very good agreement between calculation and experiment is good news for the analytical applications of computational chemistry.

SERS of thiophenol has been studied by several groups. Joo [25] has reported a surface-enhanced Raman scattering study showing that benzenethiol adsorbed faster than thiophenol on gold nanoparticle surfaces at a low bulk concentration of $\sim 10^{-7}$ M, whereas they appeared to exhibit almost identical spectral behaviors at high concentrations. Ren *et al.* [26] have reported tip-enhanced Raman spectroscopy (gold tip is used to produce SERS) of thiophenol adsorbed on Au and Pt single-crystal surfaces. The SERS spectra of thiophenol chemisorbed to gold are identical in both cases. Ren *et al.* also observed a decay in the signal that they explained in the following terms: "However, adsorbed benzenethiol is not expected to have an appropriate absorption band in the wavelength region of the excitation laser we used. Thus, the intensity decay must be attributed to the very high electromagnetic field generated in the close vicinity of the tip apex, possibly leading to the photodesorption or photodecomposition of benzenethiol quantitatively." This is also in agreement with our assumption for the selected examples in this section, where there is no resonant Raman effect at play.

The SERS of thiophenol on citrate silver colloids recorded in our laboratory is shown in Figure 4.5, where the Raman spectrum of the neat is included for easy comparison.

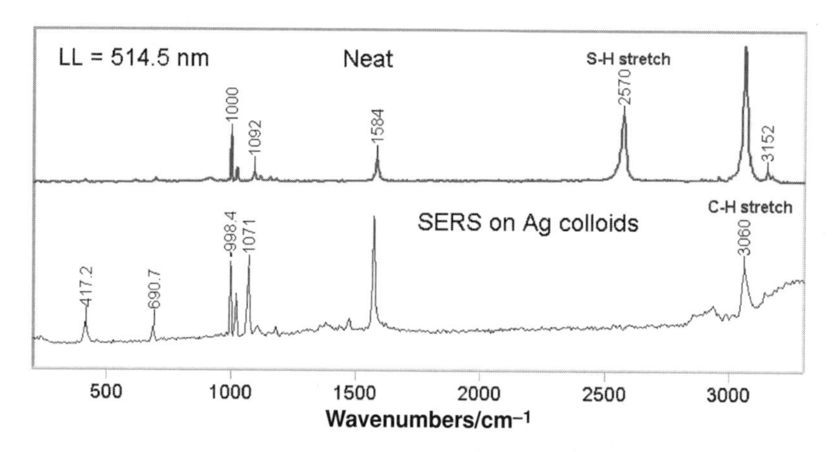

Figure 4.5 Raman spectrum of the solid and the SERS spectrum of thiophenol on citrate silver colloids. The absence of the S–H stretching mode in the SERS spectrum is compelling

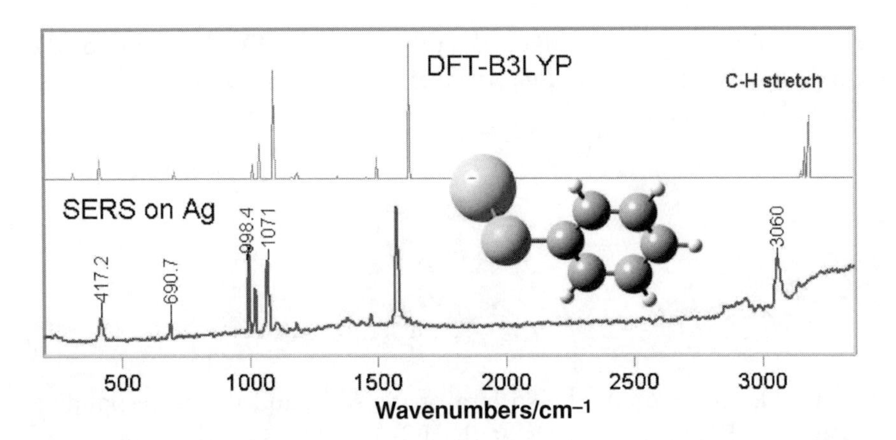

Figure 4.6 Calculated Raman spectrum of the surface complex were an Ag–S bond has been formed and the experimental SERS spectrum on silver

The SERS spectrum of thiophenol on silver is again identical with that obtained on gold nanostructures. In this case the adsorbate sends a very clear signal to the observer about the way in which it is interacting with the metal surface by the disappearance of the S–H stretching mode in the SERS spectrum and indicating that an Ag–S bond has been formed. The immediate action is to complete step 4 in our protocol and carry out the computational work for the surface complex. The results of this calculation are shown in Figure 4.6.

Again the computed spectrum is in extremely good agreement with the observed SERS spectrum, giving further support to the systematic approach as set out in the protocol of five basic steps. Clearly, very few adsorbates are as cooperative in the interpretation of the SERS spectrum as is thiophenol, and the identification of the surface complex could become the main hurdle in the spectral interpretation. This happens when there are competitive sites for adsorption within the same molecule, or when aggressive photochemistry takes place and the final product producing the spectra is only partially related to the initial structure of the target molecule.

The second study case for these classes of chemisorbed molecules is *1,8-naphthalimide*, a planar molecule which belongs to the C_{2v} point group symmetry. Following the recommendations of the report on notation for the spectra of polyatomic molecules, the *x*-axis is chosen perpendicular to the molecular plane; thereby, the molecular plane is the *yz* plane and the twofold symmetry axis is along the *z*-axis. The total irreducible representation for the 60 fundamental vibrational modes is $\Gamma = 21a_1 + 8a_2 + 20b_2 + 11b_1$, where the all the modes are Raman active with $a_1(x^2, y^2, z^2)$, $a_2(xy)$, $b_2(yz)$ and $b_1(xz)$. In the infrared, the transitions are polarized along the axes: $a_1(z)$, $b_2(y)$, $b_1(x)$, and the $8a_2$ modes are silent.

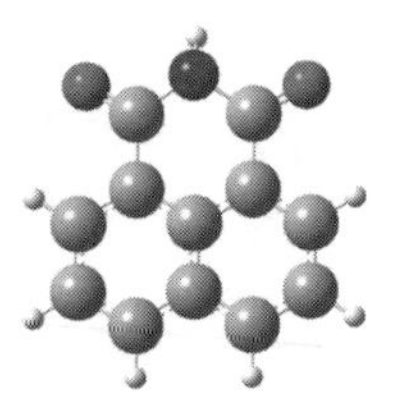

First, let us examine the spectra of the solid and the corresponding calculated Raman intensities at the B3LYP/6–311G(d) level of theory (unscaled) presented in Figure 4.7. It can be seen that the agreement is surprisingly good, given that the experimental Raman spectrum is that of the solid. The carbonyl stretching C=O vibrations in the computed spectrum appear with a large relative intensity because the symmetric and antisymmetric stretching are only one wavenumber apart, and thereby their intensities are collapsed into a single band. As expected, owing to intermolecular interactions, the C=O stretches are observed at lower

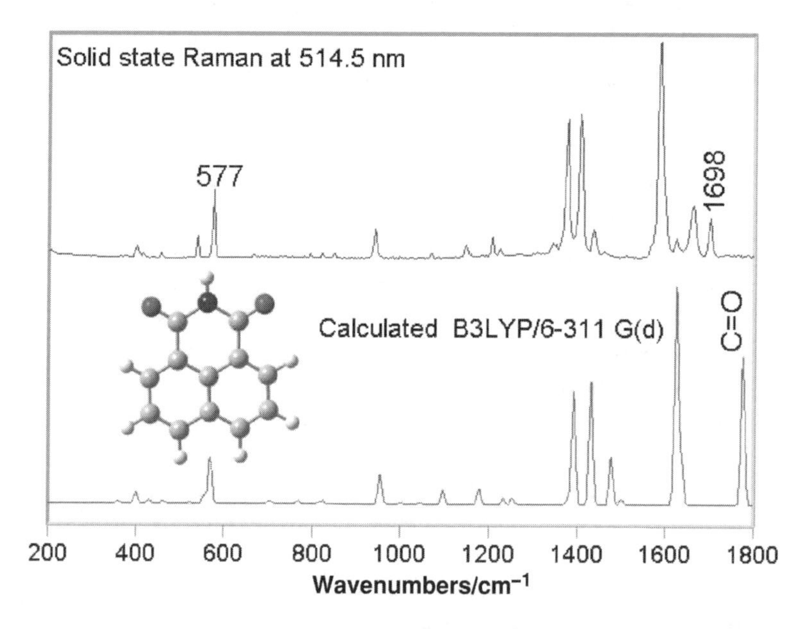

Figure 4.7 Raman scattering spectrum of solid 1,8-naphthalimide and the corresponding calculated Raman intensities and unscaled wavenumbers obtained at the B3LYP/6–311G(d) level of theory

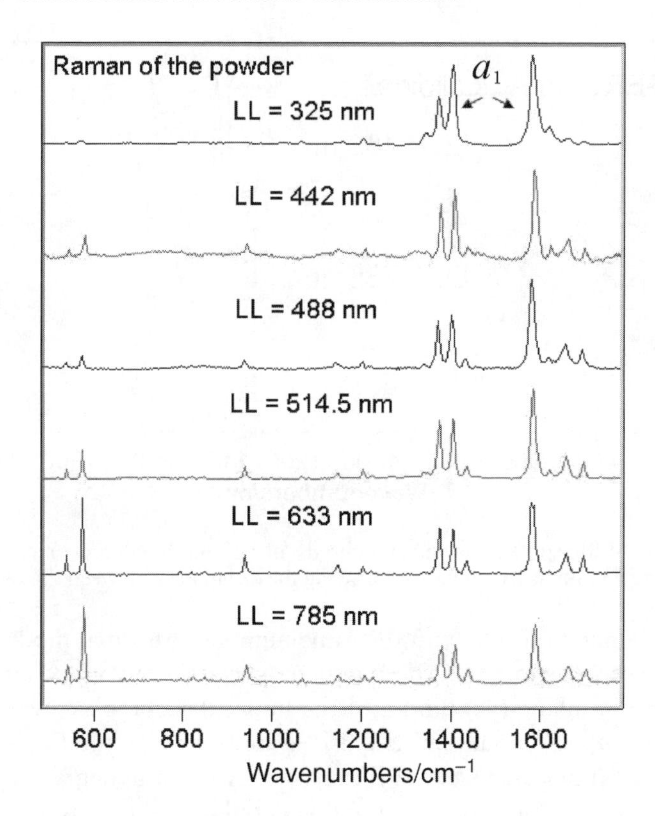

Figure 4.8 Raman and pre-resonance Raman spectra of 1,8-naphthalimide with excitation frequencies in the 325–785 nm range

wavenumber and the symmetric and antisymmetric carbonyl bands are well separated. The calculated spectrum predicts all frequencies fairly accurately, facilitating the assignment of the fundamental vibrational modes.

The excitation spectra of 1,8-naphthalimide are illustrated in Figure 4.8 for excitation frequencies in the 325–785 nm range. The laser line at 325 nm is close to resonance and it can be seen that the spectrum becomes dominated by the ring stretching modes of the chromophore in the 1400–1600 cm^{-1}. The carbonyl stretching vibrations and the deformation modes below 600 cm^{-1} lose their relative intensity approaching resonance. Off-resonance, the relative intensities are illustrated by the spectra excited with 633 or 785 nm laser radiation.

The SERS spectra were recorded on Ag colloids. It is assumed that 1,8-naphthalimide forms a surface complex by replacing the proton on the imide group by Ag. The complex preserved the C_{2v} point group symmetry

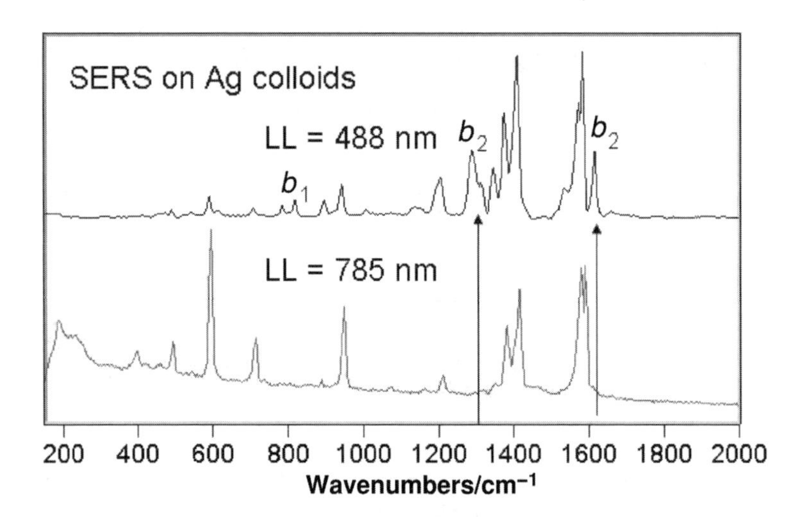

Figure 4.9 SERS spectra on silver colloids of 1,8-naphthalimide recorded with 488 nm and 785 nm laser excitation showing the differences between the two spectra

of the original molecule with the same number of normal modes and irreducible representations. The symmetry species $a_1(x^2,y^2,z^2)$ and $b_2(yz)$ are mainly in-plane (yz) modes of the molecular complex, whereas the $a_2(xy)$ and $b_1(xz)$ modes are exclusively out-of-plane modes. The local field polarization can be seen as the result of two components: E_n, perpendicular to the surface and E_{tg}, the tangential component to the surface. The SERS spectra recorded with 488 and 785 nm laser radiation are given in Figure 4.9. The most striking aspect of the experimental results is that the SERS spectrum obtained at 488 nm is busier than that obtained at 785 nm. The pattern observed in the representative 785 nm spectrum is already seen with 633 nm laser excitation, and correspondingly the busy spectrum at 488 nm is also seen when excited at 514.5 nm. The SERS spectrum excited at 633 and 785 nm is completely dominated by the a_1 species, i.e. it is a spectrum of the totally symmetric species. The same spectrum has been confirmed with excitation at 647 nm [27].

For chemisorbed 1,8-naphthalimide forming an Ag complex, we assume a head-on molecular orientation, where the z-axis of the surface Cartesian coordinates run along the $C_2(z)$ of the surface complex. The tangential component can be set parallel to the x-axis. The ratio of the two field components favors the perpendicular component towards the red, while the tangential component increases with frequency. Therefore, for an ideal head-on $[C_2(z)]$ complex and excitation to the red of 633 nm, the $a_1(z^2)$ are the most active and they dominate the spectrum. Excitation to the blue of 514.5 nm, where there is a substantial tangential

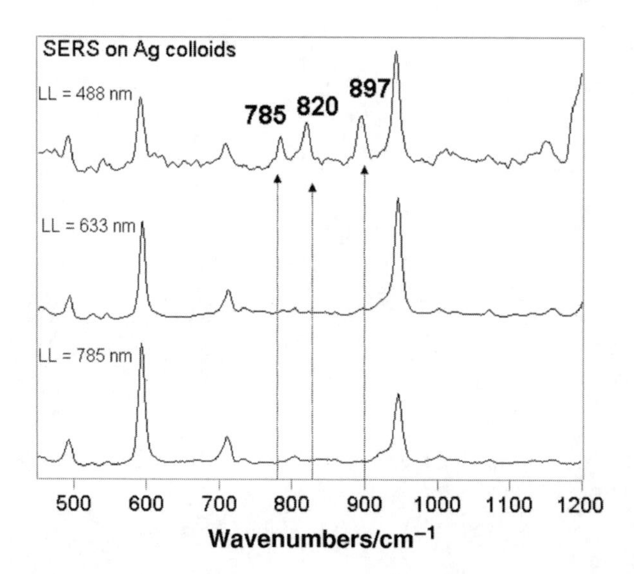

Figure 4.10 Section of the SERS spectrum of 1,8-naphthalimide on silver, highlighting the differences found with laser excitation at 488, 633 and 785 nm

component of the field at the surface, leads to a SERS spectrum that may contain all the Raman-active modes. In Figure 4.9 the bands associated with symmetry species other than the a_1 are marked. Most notably, the out-of-plane C–H wagging modes (b_1) are clearly seen in the SERS spectrum obtained with 488 nm excitation, and this section of the total SERS spectrum is shown separately in Figure 4.10. In the infrared spectrum of 1,8-naphthalimide, the C–H wagging vibrations are observed at 798 and 825 cm^{-1}. The latter modes are b_1 species and the band at 897 cm^{-1} is likely to be an a_2 mode.

The observed SERS spectra can be readily explained using the fixed orientation of the surface complex, as illustrated in Figure 4.11, and accepting that the tangential component of the electric field grows at higher frequencies of excitation [28]. The alternative explanation would be to postulate a head-on orientation of the complex, when excited with red frequencies and a mix of orientations (head-on and face-on) for high frequencies (green and blue excitation). The latter explanation is clearly a stretch of the imagination, and very unlikely to be observed. The 1,8-naphthalimide study case is a very useful example to illustrate the application of the surface selection rules taking into account the constraints introduce by the magnitude of perpendicular and the tangential components of the local electric field that may change with the excitation frequency.

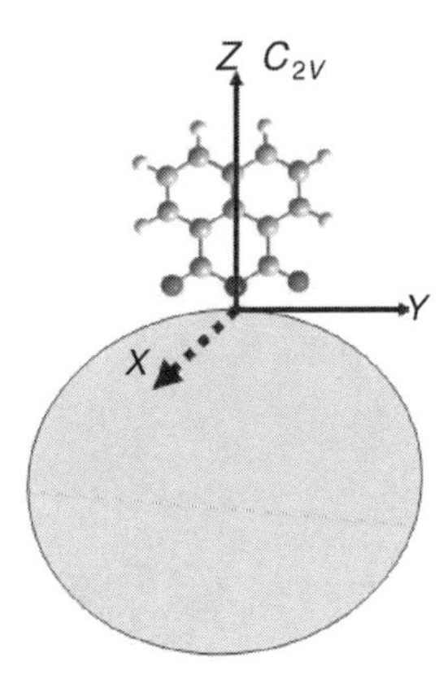

Figure 4.11 Cartoon showing the symmetry of the adsorbed target molecule of 1,8-naphthalimide

The interpretation of SERS spectra aided by quantum chemical computations is now as common as it is the assignment of vibrational spectra using quantum chemistry [21,29–31].

There has been an ongoing effort to include the enhancement effects in a systematic approach to SERS. For example, Corni and Tomasi [32,33] presented a methodology for computation of the SERS spectrum of adsorbates on metal particles possibly immersed in a solvent. The surface complex is treated *ab initio* whereas the metal particles and the solvent are described through their dielectric properties. SERS enhancement factors and Raman spectra were reported. The theoretical evaluation of the enhancement for pyridine on silver aggregates yielded *EF* values of 10^{10}.

4.4 SERS OF CHEMICALLY ADSORBED MOLECULES WITH CHARGE TRANSFER EXCITATION

CT excitation has occupied a prominent position in the theory and experiment of SERS. The book on selected SERS papers by Kerker [34] has a section (pp. 638–692) dedicated to the CT mechanism. It is a subset of the SERS of chemisorbed species. From the outset it is accurate to state that all the CT mechanisms discussed in the literature are variations of the resonance Raman process, and thereby they are part of the SERRS group. Of the several CT mechanisms discussed in the papers in

Kerker's compilation and many other publications [1,6,35–37], they can be discussed in two separate groups:

1. The first general case of CT phenomena involves the transfer of an electron from the Fermi level of the metal to an unoccupied molecular orbital of the adsorbate or vice versa [38], as illustrated in Figure 4.12. The evidence for this particular type of excitation has been mainly obtained from electrochemical experiments [39,40].
2. The second approach of CT is the production of resonance Raman scattering from metal complexes formed with the adsorbed molecule.

Notably, these metal complexes can be part of the metal nanoparticles supporting dipole particle plasmons, or isolated complexes in the absence of nanoparticles supporting dipole particle plasmons. For instance, Etchegoin *et al.* [41] have argued that macromolecules in contact with both metallic colloids and oxygen may display unusually large hot spots in SERS enhancement, and that this effect could be attributed to resonant CT interactions (mediated by oxygen) between the surface plasmons and the molecules. The latter is an example of CT on nanoparticles supporting surface plasmons. In contrast, obtaining RRS from organic–silver nanoclusters (also referred to as silver active sites) that do not support surface plasmons has been part of the SERS literature from the very beginning and extensive discussion of this work can be found in Otto's reviews [1,5]. Again, most of the ground work has been carried out on silver, where characterization of the atomic silver layer, clusters and the surface complex has been done by several groups [17,42–45].

Wu *et al.* [43] published a detailed study of the CT states for pyridine–metal clusters (Cu, Ag and Au) summarizing at the same time the electronic properties of the small metal clusters M_n ($n = 2$–4). They calculated that the Py–Ag_2 complex has an electronic transition at 467.9 nm, whereas that of the Ag_2 dimer is at 486 nm, both in the frequency range of the laser lines produced by an argon ion laser. In summary, there is irrefutable evidence of RRS in the spectral region where most dispersive Raman system operate (visible excitation) from organic–metal complexes with the same metals that are also SERS active. For instance, Miragliota and Furtak [46] probed a single monolayer of silver that does not form silver nanoparticles as enhancers of Raman scattering. They reported enhancement due to the resonant Raman effect produced by the excitation in resonance with the electronic transition of the pyridine–silver complex. Furtak's group presented evidence for Ag cluster vibrations in enhanced Raman scattering from an Ag surface, and provided an identification of these cluster as being Ag_4^+. More intriguing are recent reports where the claim is made that nanoclusters of a few atoms of silver (2–8 Ag atoms) can be used for single molecule detection (SMD) of organics directly attached to them [47]. The latter implies the achievement of a RRS cross-section much higher (by seven order of magnitude) than the $\sim 10^{-24}$ cm^2 cross-section known for the best RRS molecules such as β-carotene (2.2×10^{-24} cm^2 molecule^{-1} sr^{-1} measured for the 1520 cm^{-1} Raman band). The latter SMD results will have to be confirmed by independent measurements.

4.5 METAL–MOLECULE OR MOLECULE–METAL CHARGE TRANSFER

Laser excitation brings about the possibility of a CT transition between the Fermi level of the metal nanostructure and the LUMO of the adsorbed molecule, metal-to-molecule or molecule-to-metal transfer [36,38,40]. The latter is illustrated in Figure 4.12, where *IP* is the ionization potential of the adsorbate, φ is the work function of the metal, and the Fermi energy level is between the HOMO and the LUMO energies of the adsorbate. For CT to take place, the molecule must be chemisorbed on the metal substrate. The incoming laser line could be in resonance with the electronic transition (for instance, metal–molecule transition), producing a case of RRS. The laser line could also be in resonance with the plasmon absorption of the surface nanostructures, giving rise to electromagnetic

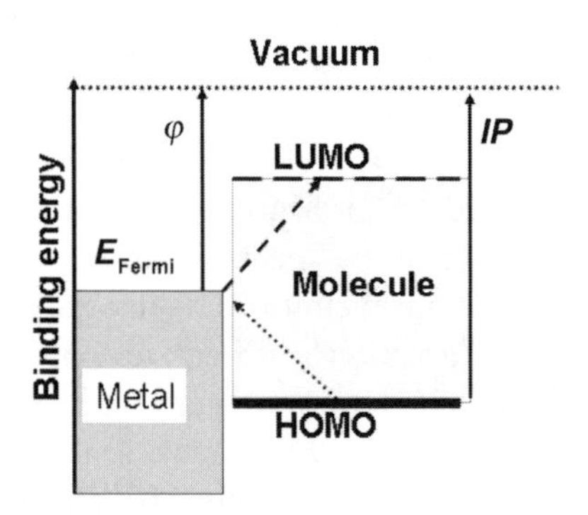

Figure 4.12 Energy diagram illustrating the band energy of the metal nanostructure and the HOMO–LUMO gap of the adsorbed molecule

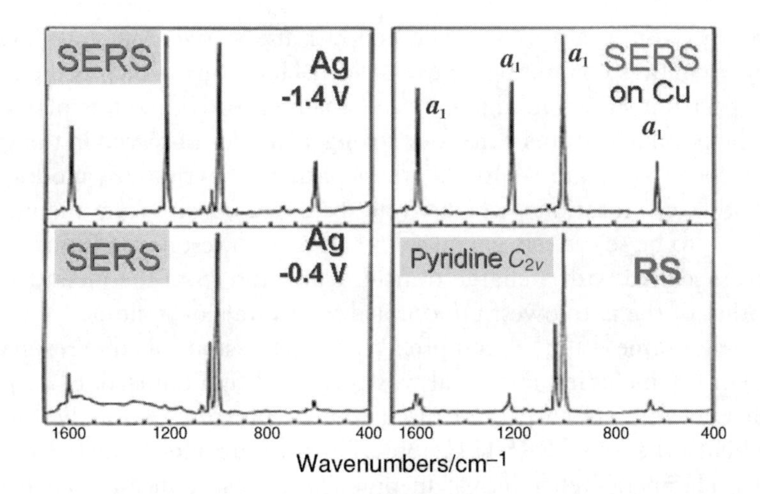

Figure 4.13 Spectroscopic results, taken from Creighton's work and rearranged, to illustrate the observation of the charge transfer for pyridine on copper or silver enhancing surfaces [49]

enhancement and the observation of SERRS. Clearly, the effect is limited to the first layer of adsorbed molecules – a *first-layer effect*. The molecule may be chemically adsorbed on the metal particle, thereby forming a new chemical bond that would be observable in the SERRS spectrum. Very early it was proposed that "charge-transfer excitations between the metal and the adsorbed molecules can give rise to an enhancement of $\sim 10^2$" [48]; however, the RRS contribution to SERRS could much higher than 10^2.

An excellent example of the observation of CT was reported by Creighton [49]. Spectroscopic results taken from Creighton's work are shown in Figure 4.13.

The inelastic Raman scattering (RS) of a 20% aqueous pyridine solution excited with 647 nm laser radiation is shown as the reference spectrum. The SERS spectrum of chemisorbed pyridine on Ag roughened electrode recorded at -0.4 V is similar to the neat RS spectrum, although there are changes (frequency shifts) due to the chemisorption on the silver surface. The top spectra are the SERS spectra of the same pyridine chemisorbed on the silver electrode when the potential is -1.4 V and on a Cu colloid. At the negative potential of the Ag electrode, a CT contribution seems to be possible, and correspondingly the spectrum shows the RRS contribution that can be extracted from the intensities observed for the a_1 species of symmetry. This spectrum is indeed a surface-enhanced resonance Raman spectrum and should be labeled SERRS. Notably, the same spectrum is also observed on copper at negative potentials and on copper colloids containing an excess of $[BH_4]^-$ ions, which is the other top spectrum shown in Figure 4.13. Creighton provides an explanation for the peculiar nature of the four strong a_1 modes observed in the spectra at the top of Figure 4.13: "... these particular pyridine a_1 modes, but not other a_1 vibrations, which would be expected on group theoretical grounds to be selectively enhanced by an *A*-term resonance Raman process associated with a charge transfer transition from the metal to one or other of the two lowest unoccupied π^* orbitals of pyridine."

In the same year, a comprehensive discussion of the resonance Raman CT including the metal was presented by Lombardi *et al.* [38], with relevant references and discussion of previous work. Following Albrecht's theory of RRS [11], the CT theory includes Franck–Condon and a Herzberg–Teller terms. In practical terms, it means that in CT SERRS spectra one could expect to see fundamental and overtones (*A* term), and also symmetric and antisymmetric modes (*B* term).

About 10 years later (1995), the SERRS theory as applied to CT was revisited by Rubim *et al.* [40] in a report where the key is the formulation of the same problem in terms of the time-dependent formalism. A complete review of this particular approach can be found in a paper by Myers [13]. The time-dependent approach was successfully used to explain CT SERRS spectra of pyridine and $[F_2(CN)_{10}bipy]^{6-}$ (bipy = 4,4′-bipyridine) on electrode surfaces and, in particular, the maximization of SERRS intensity as a function of the applied potential. An attempt is made to separate the RRS contribution from the plasmon-assisted electromagnetic enhancement.

At about the same time (1996), Oteros's group at the University of Malaga started to analyze CT SERRS spectra using their own approach based on the RRS theory. The group looked, for instance, at CT processes in surface-enhanced Raman scattering and Franck–Condon active vibrations of pyrazine [50], studying the spectroelectrochemistry of pyrazine on an Ag electrode producing SERRS spectra with relative intensities that could be explained assuming a CT effect and using resonance Raman selection rules.

The basic approach is to postulate the formation of a surface complex (metal–molecule) between the metal and the adsorbate [39]. Under the appropriate conditions (see Figure 4.12), the incident light can bring about the transfer of an electron from the Fermi level of the metal to vacant orbitals of the adsorbate, resulting in the excited CT state. The emission, an electron–hole recombination, would give a Raman-shifted frequency whenever the molecule remains vibrationally excited. The process can be drastically simplified by reducing the resonant CT process to an excitation from the ground electronic state of the neutral molecule to that of its radical anion. Since this CT mechanism is a particular RRS process, the selective enhancement of the fundamentals can be correlated with differences between the potential energy surfaces of the electronic states involved. Therefore, the strongest CT SERRS bands coincide with those assigned to the normal modes connecting the equilibrium geometry of the neutral molecule and that of the radical anion.

The extensive experimental evidence supports the hypothesis that the CT is a special case of resonant Raman scattering. That is to say, CT excitations (either from the metal to the molecule or vice versa) with an electronic energy much lower than that of the intramolecular HOMO–LUMO excitations may be in resonance with one of the laser lines used in most dispersive Raman instruments [36,51].

The role of the resonant Raman effect has come full circle; it was hinted at as part of the observed enhanced signal in both seminal papers in 1977 by Jeanmarie and Van Duyne [7] and in the report by Albrecht and Creighton [8] (see Chapter 3). Today, the resonance Raman model is being applied in the interpretation of the observed SERS spectra under the general umbrella of CT SERRS.

4.6 SERRS FROM A SURFACE COMPLEX

The treatment of surface complex formation with electronic transitions in resonance with the excitation laser line forms part of a large body of experimental data that is also discussed as part of CT phenomena. This CT nomenclature is by analogy with the CT complexes studied in inorganic chemistry. "A charge transfer complex is one in which a donor and acceptor interact together with some transfer of electronic charge, usually facilitated by the acceptor" [52]. The following situations must be considered: full resonance with the plasmon absorption (maximum EM enhancement) and full resonance with the complex electronic absorption. For spectral interpretation, the latter will be the case in surface-enhanced resonance Raman scattering. When the laser excitation is not in full resonance with the electronic absorption of the surface complex, one can have a case of surface-enhanced pre-resonant Raman scattering. Since the absorption spectra of the surface complexes are usually not known, the interpretation of the spectra could become a guessing game. Therefore, the characterization of the surface complex is at the center of these investigations and can reveal surface complexes without electronic transitions as discussed in Section 4.3, or complexes with electronic transitions in the visible region (the spectral region where the Raman experiments are commonly carried out). The latter case is briefly discussed and illustrated here.

The main task in the characterization of the complex is to obtain its absorption spectrum, so that the resonance conditions are clearly identified. Therefore, the first task in this line of work is to record the absorption spectrum followed by the Raman excitation profiles. Without this information, the spectral interpretation provided is always questionable. If the resonant Raman effect can be ruled out, then the methodology described in Section 4.3 applies. Data in this respect are not abundant, given the difficult task of finding the absorption of surface complexes; but the effort has been there from the very beginning [53,54] and continues to be an important part of SERS experimental work [36,45,55,56], together with the computational approach to obtain information about the electronic structure of complexes [44,57].

Finally, one should discuss the RRS of CT complexes formed on surfaces that do not support surface plasmons or, more recently, the formation organic–silver complexes where the silver is in the form of a nanocluster (2–8 atoms). Metal clusters with <20 atoms are termed very small clusters, and their properties can be found in Kreibig and Vollmer's book [58]. In a recent report, Peyser-Capadona *et al.* [47] claimed that "In the absence of large, plasmon-supporting nanoparticles, biocompatible dendrimer- and peptide-encapsulated few-atom Ag nanoclusters produce scaffold-specific single molecule (SM) Stokes and anti-Stokes Raman scattering." The few atoms are Ag_n (with $n = 2$–8). The origin of the reported Raman signal will have to be confirmed independently. An independent confirmation of these findings will give rise to super-resonance Raman scattering, i.e. a Raman cross-section for silver–organic complexes that would be at least eight orders of magnitude better than the best known resonance Raman scatterer. The cross-sections for pyridine–Ag_2 [57] and small pyridine–Ag_n ($n = 2$–4) aggregates have been calculated [43]. The latter findings do not point to extraordinary cross-sections. In an *ab initio* study of the absorption spectra of Ag_n ($n = 5$–8) clusters [59], the calculated absorption for Ag_5, Ag_7 and Ag_8 was in agreement with experimental observations in solid argon with the most intense peaks seen below 350 nm. Notably, the findings regarding the emission properties were very interesting: "An important aspect of optical properties of silver clusters is their ability to fluoresce, which indicates that the lifetime of the dominant resonance is not extremely short. Our preliminary calculations of geometry relaxation in the excited states of Ag_8 cluster confirm this finding. Moreover, this phenomenon can be enhanced through the doping of silver clusters by oxygen atom which can activate the excitation of d-electrons in Ag atoms of the clusters giving rise to blinking as recently observed." The fluorescence properties were realized in water-soluble dendrimer-encapsulated silver clusters [60]. In summary, the RRS of a CT complex should be treated as such, a case of resonance Raman not SERS. The plasmon-assisted secondary emission, resonance Raman scattering of a CT complex, is a genuine case of SERRS.

REFERENCES

[1] A. Otto, The chemical (electronic) contribution to SERS, *J. Raman Spectrosc.* 2005; **36**: 497–509.
[2] D.M. Ruthven, *Principles of Adsorption and Adsorption Processes*, John Wiley & Sons, Inc., New York, 1984.
[3] A.W. Adamsom, *Physical Chemistry of Surfaces*, John Wiley & Sons, Inc., New York, 1990.
[4] J.C. Decius and R.M. Hexter, *Molecular Vibrations in Crystals*, McGraw-Hill, New York, 1977.
[5] A. Otto, I. Mrozek, H. Grabhorn and W. Akemann, Surface-enhanced Raman scattering *J. Phys.Condens. Matter* 1992, **4**, 1143–1212.
[6] M.E. Lippitsch, Ground-state charge transfer as a mechanism for SERS, *Phys. Rev. B* 1984, **29**, 3101–3110.
[7] D.L. Jeanmaire and R.P. Van Duyne, Surface Raman spectroelectrochemistry. Part I. Heterocyclic, aromatic, and aliphatic amines adsorbed on the anodized silver electrode *J. Electroanal. Chem.* 1977, **84**, 1–20.
[8] M.G. Albrecht and J.A. Creighton, Anomalously intense Raman spectra of pyridine at a silver electrode *J. Am. Chem. Soc.* 1977, **99**, 5215–5217.
[9] J. Behringer, in H.A. Szymanski (ed.), *Raman Spectroscopy*, Plenum Press, New York, 1967, Chapt. 6.
[10] P.P. Shorygin and L.L. Krushinskij, Early days and later development of resonance Raman spectroscopy, *J. Raman Spectrosc.* 1997, **28**, 383–388.
[11] J. Tang and A.C. Albrecht, in H.A. Szymanski (ed.), *Raman Spectroscopy*, Plenum Press, New York, 1970, Chapt. 2.
[12] D.L. Rousseau, J.M. Friedman and P.F. Williams, The resonance Raman effect *Top. Curr. Phys.* 1979, **11**, 203–252.
[13] A.B. Myers, 'Time-dependent' resonance Raman theory, *J. Raman Spectrosc.* 1997, **28**, 389–401.
[14] P.J.G. Goulet, N.P.W. Pieczonka and R.F. Aroca, Overtones and combinations in single-molecule surfaced-enhance resonance Raman scattering spectra, *Anal. Chem.* 2003, **75**, 1918–1923.
[15] R.O. Loutfy and R. Aroca, Interaction of indium metal with phthalocyanine molecules: luminescence enhancement, *J. Lumin.* 1982, **26**, 359–366.

[16] C.S. Allen and R.P. Van Duyne, Molecular generality of surface-enhanced Raman spectroscopy (SERS). A detailed investigation of the hexacyanoruthenate ion adsorbed on silver and copper electrodes. *J. Am. Chem. Soc.* 1981, **103**, 7497–7501.
[17] D. Roy and T.E. Furtak, Characterization of surface complexes in enhanced Raman scattering, *J. Chem. Phys.* 1984, **81**, 4168–4175.
[18] M. Moskovits, Surface selection rules *J. Chem. Phys.* 1982, **77**, 4406–4416.
[19] W.-H. Yang, J. Hulteen, G.C. Schatz and R.P. Van Duyne, A surface-enhanced hyper-Raman and surface-enhanced Raman scattering study of *trans*-1,2-bis(4-pyridyl)ethylene adsorbed onto silver film over nanosphere electrodes. Vibrational assignments: experiment and theory, *J. Chem. Phys.* 1996, **104**, 4313–4323.
[20] R.F. Aroca, R.E. Clavijo, M.D. Halls and H.B. Schlegel, Surface-enhanced Raman spectra of phthalimide. Interpretation of the SERS spectra of the surface complex formed on silver islands and colloids, *J. Phys. Chem. A* 2000, **104**, 9500–9505.
[21] T. Tanaka, A. Nakajima, A. Watanabe, T. Ohno and Y. Ozaki, Surface-enhanced Raman scattering of pyridine and *p*-nitrophenol studied by density functional theory calculations, *Vib. Spectrosc.* 2004, **34**, 157–167.
[22] G. Cardini and M. Muniz-Miranda, Density functional study on the adsorption of pyrazole onto silver colloidal particles, *J. Phys. Chem. B* 2002, **106**, 6875–6880.
[23] M.D. Halls, J. Velkovski and H.B. Schlegel, Harmonic frequency scaling factors for Hartree–Fock, S-VWN, B-LYP, B3-LYP, B3-PW91 and MP2 with the Sadlej pVTZ electric property basis set, *Theor. Chem. Acc.* 2001, **105**, 413–421.
[24] A.P. Scott and L. Radom, Harmonic Vibrational Frequencies: An Evaluation of HartreeFock, Møller-Plesset, Quadratic Configuration Interaction, Density Functional Theory and Semiempirical Scale Factors, *J. Phys. Chem.* 1996, **100**, 16502–16513.
[25] S.-W. Joo, Adsorption of aromatic thiols on gold nanoparticle surfaces investigated by UV–vis absorption spectroscopy and surface enhanced Raman scattering, *Chem. Lett.* 2004, **33**, 60–61.
[26] B. Ren, G. Picardi, B. Pettinger, R. Schuster and G. Ertl, Tip-enhanced Raman spectroscopy of benzenethiol adsorbed on Au and Pt single-crystal surfaces, *Angew. Chem. Int. Ed.* 2004, **44**, 139–142.
[27] J.R. Menendez, A. Obuchowska and R. Aroca, Infrared spectra and surface enhanced Raman scattering of naphthalimide on colloidal silver, *Spectrochim. Acta, Part A* 1996, **52**, 329–336.
[28] M. Moskovits, Surface enhanced spectroscopy, *Rev. Mod. Phys.* 1985, **57**, 783–826.
[29] B. Sagmuller, P. Freunscht and S. Schneider, The assignment of the vibrations of substituted mercaptotetrazoles based on quantum chemical calculations, *J. Mol. Struct.* 1999, **482–483**, 231–235.
[30] S.W. Han, S.W. Joo, T.H. Ha, Y. Kim and K. Kim, Adsorption characteristics of anthraquinone-2-carboxylic acid on gold, *J. Phys. Chem. B* 2000, **104**, 11987–11995.
[31] P. Bleckmann, M. Thibud and H.D. Trippe, Characterization of the surface enhanced Raman scattering by use of Raman-spectroscopic and quantum mechanical investigations of simple cluster compounds, *J. Mol. Struct.* 1988, **174**, 59–64.
[32] S. Corni and J. Tomasi, Theoretical evaluation of Raman spectra and enhancement factors for a molecule adsorbed on a complex-shaped metal particle, *Chem. Phys. Lett.* 2001, **342**, 135–140.

[33] S. Corni and J. Tomasi, Surface enhanced Raman scattering from a single molecule adsorbed on a metal particle aggregate. A theoretical study, *J. Chem. Phys.* 2002, **116**, 1156–1164.

[34] M. Kerker, (ed), *Selected Papers on Surface-enhanced Raman Scattering*, SPIE, Bellingham, WA, 1990.

[35] T.E. Furtak, Current understanding of the mechanism of surface enhanced Raman scattering. *J. Electroanal. Chem. Interfacial Electrochem.* 1983, **150**, 375–388.

[36] A. Campion and P. Kambhampati, Surface-enhanced Raman scattering, *Chem. Soc. Rev.* 1998, **27**, 241.

[37] J.F. Arenas, J. Soto, I. Lopez Tocon, D.J. Fernandez, J.C. Otero and J.I. Marcos, The role of charge-transfer states of the metal–adsorbate complex in surface-enhanced Raman scattering, *J. Chem. Phys.* 2002, **116**, 7207–7216.

[38] J.R. Lombardi, R.L. Birke, T. Lu and J. Xu, Charge-transfer theory of surface enhanced Raman spectroscopy: Herzberg–Teller contributions, *J. Chem. Phys.* 1986, **84**, 4174–4180.

[39] J.F. Arenas, D.J. Fernandez, J. Soto, I. Lopez-Tocon and J.C. Otero, Role of the electrode potential in the charge-transfer mechanism of surface-enhanced Raman scattering, *J. Phys. Chem. B* 2003, **107**, 13143–13149.

[40] J.C. Rubim, P. Corio, M.C.C. Ribeiro and M. Matz, Contribution of resonance Raman scattering to the surface-enhanced Raman effect on electrode surfaces. A description using the time dependent formalism, *J. Phys. Chem.* 1995, **99**, 15765–15774.

[41] P. Etchegoin, H. Liem, R.C. Maher, L.F. Cohen, R.J.C. Brown, H. Hartigan, M.J.T. Milton and J.C. Gallop, A novel amplification mechanism for surface enhanced Raman scattering, *Chem. Phys. Lett.* 2002, **366**, 115–121.

[42] D. Roy and T.E. Furtak, Evidence for silver cluster vibrations in enhanced Raman scattering from the silver/electrolyte interface, *Chem. Phys. Lett.* 1986, **124**, 299–303.

[43] D.Y. Wu, M. Hayashi, C.H. Chang, K.K. Liang and S.H. Lin, Bonding interaction, low-lying states and excited charge-transfer states of pyridine–metal clusters: pyridine–M_n ($M = Cu, Ag, Au$; $n = 2$–4), *J. Chem. Phys.* 2003, **118**, 4073–4085.

[44] I.S. Alaverdian, A.V. Feofanov, S.P. Gromov, A.I. Vedernikov, N. ya. Lobova and M.V. Alfimov, Structure of charge-transfer complexes formed by biscrown stilbene and dipyridylethylene derivatives as probed by surface-enhanced Raman scattering spectroscopy, *J. Phys. Chem. A* 2003, **107**, 9542–9546.

[45] I. Srnova-Sloufova, B. Vlckova, T.L. Snoeck, D.J. Stufkens and P. Matejka, Surface-enhanced Raman scattering and surface-enhanced resonance Raman scattering excitation profiles of Ag–2,2′-bipyridine surface complexes and of $[Ru(bpy)_3]^{2+}$ on Ag colloidal surfaces: manifestations of the charge-transfer resonance contributions to the overall surface enhancement of Raman scattering, *Inorg. Chem.* 2000, **39**, 3551–3559.

[46] J. Miragliotta and T.E. Furtak, Enhanced Raman scattering with one monolayer of silver, *Phys. Rev. B* 1987, **35**, 7382–7391.

[47] L. Peyser-Capadona, J. Zheng, J.I. Gonzalez, T.-H. Lee, S.A. Patel and R.M. Dickson, Nanoparticle-free single molecule anti-stokes Raman spectroscopy, *Phys. Rev. Lett.* 2005, **94**, 58301–58304.

[48] B.N.J. Persson, On the theory of surface-enhanced Raman scattering, *Phys. Lett.* 1981, **82**, 561–565.

[49] J.A. Creighton, The resonance Raman contribution to SERS: pyridine on copper or silver in aqueous media, *Surf. Sci,* 1986, **173**, 665–672.

[50] J.F. Arenas, M.S. Woolley, J.C. Otero and J.I. Marcos, Charge-transfer processes in surface-enhanced Raman scattering. Franck–Condon active vibrations of pyrazine, *J. Phys. Chem.* 1996, **100**, 3199–3206.

[51] A.G. Brolo, D.E. Irish, G. Szymanski and J. Lipkowski, Relationship between SERS Intensity and both surface coverage and morphology for pyrazine Adsorbed on a polycrystalline gold electrode, *Langmuir* 1998, **14**, 517–527.

[52] C.E. Housecroft and A.G. Sharpe, *Inorganic Chemistry*, Pearson Education, Harlow, 2005.

[53] D.A. Weitz, S. Garoff and T.J. Gramila, Excitation spectra of surface-enhanced Raman scattering on silver-island films, *Opt. Lett.* 1982, 7, 168–170.

[54] I. Pockrand, J. Billmann and A. Otto, Surface enhanced Raman scattering (SERS) from pyridine on silver–UHV interfaces: excitation spectra, *J. Chem. Phys.* 1983, **78**, 6384–6390.

[55] V. Oklejas and J.M. Harris, Potential-dependent surface-enhanced Raman scattering from adsorbed thiocyanate for characterizing silver surfaces with improved reproducibility, *Appl. Spectrosc.* 2004, **58**, 945–951.

[56] K. Shibamoto, K. Katayama and T. Sawada, Fundamental processes of surface enhanced Raman scattering detected by transient reflecting grating spectroscopy, *J. Photochem. Photobiol., A* 2003, **158**, 105–110.

[57] D.Y. Wu, M. Hayashi, S.H. Lin and Z.Q. Tian, Theoretical differential Raman scattering cross-sections of totally-symmetric vibrational modes of free pyridine and pyridine–metal cluster complexes, *Spectrochim. Acta, Part A* 2004, **60**, 137–146.

[58] U. Kreibig and M. Vollmer, *Optical Properties of Metal Clusters*, Springer-Verlag, Berlin, 1995.

[59] V. Bonacic-Koutecky, V. Veyret and R. Mitric, *Ab initio* study of the absorption spectra of Ag_n (n = 5–8) clusters, *J. Chem. Phys.* 2001, **115**, 10450–10460.

[60] J. Zheng and R.M. Dickson, Individual water-soluble dendrimer-encapsulated silver nanodot fluorescence, *J. Am. Chem. Soc.* 2002, **124**, 13982–13983.

5

Is SERS Molecule Specific?

The observation of giant intensities in SERS has always been confronted with the fact that some molecules apparently do not show enhancement factors expected for substrates excited in resonance with their plasmon absorption bands. Some classical examples of this include water, methanol and alkanes in general. As the observed experimental intensity for any SERS spectrum is simply the 'tip of the iceberg', one does not get the benefit of distinguishing the real contributions to the vibrational intensity. In short, if one enhances a vibrational intensity with an absolute value of 0.1 by 10^6-fold, that signal will be 10^3 times weaker than a vibrational intensity with an initial absolute magnitude of 100, and the first signal will not make it out of the noise, whereas the second will have a very healthy signal-to-noise ratio. Instead of absolute intensities (see Chapter 1), the quantity that allows one to compare the efficiencies of the optical processes is the cross-section and a few examples may be of help in appreciating the context within which enhancement factors are observed. In spontaneous inelastic Raman scattering (RS), the total Stokes scattered light, averaged over all random molecular orientations, I_{RS} (photons s^{-1}) is proportional to the incoming flux of photons, I_0 (photons s^{-1} cm^{-2}):

$$I_{RS} = \sigma_{RS} I_0. \tag{5.1}$$

The proportionality constant, the Raman cross-section σ_{RS}, has the dimensions of cm^2 and is a function of the frequency of excitation.

First, a brief look at the molecular Raman and infrared cross-sections should help to illustrate the differences found in these quantities for a single molecule. In Table 5.1, the infrared and Raman absolute intensities

Surface-Enhanced Vibrational Spectroscopy R. Aroca

Table 5.1 Infrared and Raman absolute intensities and corresponding cross-section for the water molecule

	Infrared intensity [3][a]		Raman intensity [2][b]	
Wavenumber/cm^{-1}	km/mol^{-1}	Cross-section/ cm^2 molecule^{-1}	Å^4/amu	Cross-section (514.5 nm)/ cm^2 molecule^{-1} sr^{-1}
1595	62.5	6.5×10^{-21}	10.7	0.11×10^{-30}
3657	2.9	1.31×10^{-22}	86.1	
3756	41.7	1.8×10^{-21}	36.4	

[a] Experiment.
[b] Calculated.

and the corresponding cross-sections are summarized for the water molecule.

The infrared cross-sections were calculated from the integrated intensities Γ that are in units of cm^2 mol^{-1} [4]:

$$\Gamma = \frac{1}{cl} \int_{\text{Band}} \ln\left(\frac{I_0}{I}\right) \mathrm{d}\ln\nu. \tag{5.2}$$

It can be seen that there is a difference in the dimensions of the infrared and Raman cross-sections of about 10 orders of magnitude, which explains the widespread preferred use of infrared over Raman techniques. According to Equation (5.2), $I_{\text{RS}} = \sigma_{\text{RS}} I_0$, and the Raman sensitivity may be improved by increasing the influx of photons, which was achieved with the advent of lasers, and/or by increasing the cross-section of the scatterer. Notably, the high sensitivity of detectors in the visible region [charge-coupled device (CCD)], which permits the detection of very weak signals in scattering experiments, has already advanced spontaneous Raman spectroscopy, making it a routine analytical technique [2,5,6].

Second, we should compare the Raman cross-sections for different molecules. Table 5.2 gives some Raman cross-section values for common

Table 5.2 Raman cross-sections for common Raman scatterers

Molecule	Vibration/cm^{-1}	Excitation/nm	Raman cross-section/ cm^2/molecule^{-1}/sr^{-1}
Benzene (liquid)	3060	514.5	45.3×10^{-30}
Cyclohexane (liquid)	1444	488	6.2×10^{-30}
CH_2Cl_2 (gas)	713	514.5	2.3×10^{-30}
CCl_4 (gas)	459	514.5	4.7×10^{-30}

Raman scatterers [2]. It can be seen that these cross-sections vary from about 20 to 450 times more strongly than that of water, thus explaining why water is such a good solvent for Raman scattering studies.

Third, the frequency dependence of the cross-section (dispersion) leads to a special case when the laser excitation frequency approaches or is in resonance with the molecular electronic excited state. This gives rise to pre-resonance or resonance Raman scattering, where the polarizability derivative term contained in the cross-section is resonantly enhanced. The absolute Raman cross-sections for the 666 cm^{-1} mode of chloroform have been determined [7] using several laser lines and were reproduced in Chapter 1 (Table 1.7) to illustrate the dispersion of the cross-section for this vibrational mode as resonance is approached.

The increasing cross-section with the wavelength approaching the electronic excited state is clear. The molecular cross-section is dependent on the transition electronic polarizability tensor α, which is responsible for the scattering of visible or near-infrared light. The classical dispersion theory of dielectric media (reference 8, p. 309) gives an expression for the polarizability of one electron bound by a harmonic force that clearly rationalizes the resonance phenomena (see Chapter 2):

$$p = \alpha\, E = \frac{\frac{e^2}{m}}{{\omega_0}^2 - \omega^2 - i\gamma\omega} E \qquad (5.3)$$

where γ is the damping factor, ω_0 is the natural frequency of the oscillator, e is the electron charge and m is the electron mass. When ${\omega_0}^2 - \omega^2 = 0$ (the resonance condition), the magnitude of the polarizability increases, capped only by the damping term. Employing quantum mechanics, the electronic polarizability in the induced transition dipole moment, which is derived by second-order perturbation theory was given in Chapter 4 (4.2).

The first term in expression (4.2) gives rise to the resonance condition, and the complete development for the vibronic expansion leads to the Albrecht $A + B + C$ terms [9]. Experimentally observed resonance Raman spectra are correspondingly assigned to, or explained in terms of, one or more of these three terms. Experimentally observed intensities in resonance Raman scattering (RRS), can be illustrated with a simple example. The RRS cross-section, σ_{RRS}, for β-carotene in benzene is reported to be 1.1×10^{-23} cm^2 $molecule^{-1}$ sr^{-1} for the vibrational band at 1520 cm^{-1}, and for the 1005 cm^{-1} band σ_{RRS} is 2.2×10^{-24} [2]. The main message from the magnitudes of the RRS is that they can provide a range of eight orders of magnitude for cross-section values, as can be seen when comparing water (0.11×10^{-30}) with β-carotene ($\sim 10^{-23}$).

SERS intensities expressed in terms of an effective cross-section, σ_{SERS}, allow one to write the general form for the measured SERS intensity, $I_{RS} = \sigma_{SERS} I_0$. The ratio of the enhanced to the normal Raman cross-section is the SERS enhancement factor, $EF = \sigma_{SERS}/\sigma_{RS}$, or $EF = \sigma_{SERRS}/\sigma_{RRS}$ for SERRS. Determining the enhancement factor from experiments has been one of the objectives of SERS work, and many different experimental protocols have been proposed. The size of the enhancement factor is a key property for the analytical application of SERS in ultrasensitive chemical analysis. Therefore, even in the simplest and ideal case of a constant surface enhancement factor, and keeping the molecular density constant, the experimental SERS results may be positive for some molecules and not immediately obvious for others, because the level of detection could be dramatically different. For instance, let us assume a constant electromagnetic enhancement factor of 10^3 and keep the illuminated molecular density constant. We could have $EF \times \sigma_{RS} = \sigma_{SERS}$ ($10^3 \times 0.11 \times 10^{-31} = 1.10 \times 10^{-29}$) for water and $EF \times \sigma_{RRS} = \sigma_{SERRS}$ ($10^3 \times 1.1 \times 10^{-23} = 1.1 \times 10^{-20}$) for β-carotene. That alone could lead to the wrong conclusion that SERS is molecular specific, since it is so easy to obtain SERRS of rhodamine and β-carotene, but not that of methanol. In practice, one has to deal with an array of factors that strongly affect the observed Raman signal and its intensity. These effects have created in many instances the wrong impression about SERS. To maximize the SERS signal there are several factors at the disposal of the experimentalist that obviously depend on the type of SERS experiment, relating to, e.g., colloidal solutions, metal films or electrochemistry. However, at least two generic factors should always be optimized. The first is the plasmon resonance of the ensemble of nanoparticles used for SERS that should be in tune with the laser line used for the excitation of the Raman scattering. A laser line slightly to the red of the center of the plasmon absorption is probably the best choice, since it maximizes the participation of aggregates of particles in that distribution. The second factor in all cases is the need to increase the adsorption of the target molecule on the surface of the enhancing nanoparticle. The latter can be achieved, for example, by controlling the surface charge of the nanoparticles in solution or the voltage applied to the electrode, or eliminating surface competitors (small molecules that bind to the surface).

An attempt was made to collect in one place the information on molecules that have been studied using SERS or SERRS. Seki, in 1986, published the first collection of molecules considered to be surface enhanced [10]. From this first database, it is already clear that a wide range of neutral and ionic molecules were SERS active. The collected database offers irrefutable evidence that SERS is not limited to any particular group of molecules and, in principle, it can be apply to all molecular systems. The first item in the database is a list of keywords used to group the molecular systems that had been studied by SERS/SERRS, or keywords representing a type of SERS study (for instance, pyridine, dye, and electrochemistry). A search of the keyword "dye" on the spreadsheet will point to the "dye SERS paper" published in a given year. The title of the corresponding paper is preceded by two numbers. The first number is the number assigned to the paper, and the second number corresponds to the year of publication. The references are then listed by year of publication.

The first reported enhancement factors by Jeanmarie and Van Duyne [11] were for pyridine, although, the paper states that "... we have found that many other amines, both aliphatic and aromatic, produce intense Raman scattering when adsorbed on a silver electrode." Further, "Given that the experimentally observed intensities of the NR scattering from adsorbed pyridine in our laboratory are 5–6 orders of magnitude greater than expected, we felt that some property of the electrode surface or the electrode/solution interface is acting to enhance the effective Raman scattering cross sections for these adsorbed amines."

Albrecht and Creighton [12] also reported an enhancement factor for pyridine of 10^5: "We are thus led to conclude that there is a considerable enhancement ($\sim\times10^5$) in the spectra of adsorbed pyridine by a surface effect which greatly increases the molecular Raman scattering cross-sections." After these two seminal studies, the enhancement of pyridine on silver and other enhancing surfaces was measured many times. In the year after these reports, experiments were tuned to extract the origin of the new phenomena [13–15] and proposed theoretical models [16]. Also a few new molecules were studied [17,18].

In 1979, activity continued to be on the understanding of enhancement involving pyridine [19–24], cyanide [20,25], cyanopyridines [26], Ag–Cl vibration [27] and pyridinecarboxaldehyde [28]. Theoretical models were also developed [29–31].

A list of molecules studied using SERS or SERRS has been compiled starting in 1980.* First, a definition of the keywords used in the classification of the topics found in the literature is given. The references

---

* A comprehensive reference database of more than 3000 references from the literature from 1980 to 2005 with a listing of references by key word is provided on www.spectroscopynow.com

are given for each year up to 2005. The reported enhancement factors are also found to cover wide range of values, in agreement with the fact that every enhancing structure, given the conditions of the experiment, produces an enhancement factor specific for that set of variables. These variables, among others, include the size, shape and dielectric function of the nanostructure, influenced by surface coverage, dielectric constant of the medium, adsorption of the target molecule, metal–molecule interactions, molecular orientation and polarization effects. Nevertheless, the conclusion is that for ensemble measurements and 'average' SERS, enhancement from metal nanostructures is always possible, with the caveat that the enhancement factor may be within a fairly wide range of values, where a minimum enhancement of 10^2 can be achieved even under the conditions of FT-SERS, at 1064 nm excitation.

REFERENCES

[1] A.C. Albrecht and M.C. Hutley, On the dependence of vibrational Raman intensity on the wavelength of the incident light, *J. Chem. Phys.* 1971, **55**, 4438–4443.

[2] R.L. McCreery, *Raman Spectroscopy for Chemical Analysis*, John Wiley & Sons, Inc., New York, 2000.

[3] D.M. Bishop and L.M. Cheung, Vibrational contributions to molecular dipole polarizabilities, *J. Phys. Chem. Ref. Data* 1982, **11**, 119–133.

[4] M. Mills, Infrared intensities, *Annu. Rep. Chem. Soc. London* 1958, **55**, 55–67.

[5] M. Diem, *Modern Vibrational Spectroscopy*, John Wiley & Sons, Inc., New York, 1993.

[6] K. Nakamoto, *Infrared and Raman Spectra of Inorganic and Coordination Compounds. Part A: Theory and Applications in Inorganic Chemistry*, John Wiley & Sons, Inc., New York, 1997.

[7] C.E. Foster, B.P. Barham and P.J. Reida, Resonance Raman intensity analysis of chlorine dioxide dissolved in chloroform: the role of nonpolar solvation, *J. Chem. Phys.* 2001, **114**, 8492–8504.

[8] D.J. Jackson, *Classical Electrodynamics*, John Wiley & Sons, Inc., New York, 1999.

[9] J. Tang and A.C. Albrecht, in H.A. Szymanski (ed.), *Raman Spectroscopy*, Plenum Press, New York, 1970, Chapt. 2.

[10] H. Seki, Raman spectra of molecules considered to be surface enhanced, *J. Electron Spectrosc. Relat. Phenom.* 1986, **39**, 289–310.

[11] D.L. Jeanmaire and R.P. Van Duyne, Surface Raman spectroelectrochemistry, *J. Electroanal. Chem.* 1977, **84**, 1–20.

[12] M.G. Albrecht and J.A. Creighton, Anomalously intense Raman spectra of pyridine at a silver electrode, *J. Am. Chem. Soc.* 1977, **99**, 5215–5217.

[13] M.G. Albrecht, J.F. Evans and J.A. Creighton, The nature of an electrochemically roughened silver surface and its role in promoting anomalous Raman scattering intensity, *Surf. Sci.* 1978, **75**, L777–L780.

[14] F.W. King, R.P. Van Duyne and G.C. Schatz, Theory of Raman scattering by molecules adsorbed on electrode surfaces, *J. Chem. Phys.* 1978, **69**, 4472–4481.

[15] B. Pettinger and U. Wenning, *Raman Spectra of Pyridine Adsorbed on Silver Single Crystal Electrodes with Different Crystallographic Orientation*, Fritz-Haber-Institute, Max-Planck-Gesellschaft, Berlin, 1978, pp. 169–174.

[16] M. Moskovits, Surface roughness and the enhanced intensity of Raman scattering by molecules adsorbed on metals, *J. Chem. Phys.* 1978, **69**, 4159–4161.

[17] A. Otto, Raman spectra of cyanide ion and carboxyl adsorbed at a silver surface, *Ned. Tijdschr. Vacuumtech.* 1978, **16**, 139.

[18] F.R. Aussenegg and M.E. Lippitsch, On Raman scattering in molecular complexes involving charge transfer, *Chem. Phys. Lett.* 1978, **59**, 214–216.

[19] G. Blondeau, M. Froment, J. Zerbino, N. Jaffrezic-Renault and G. Revel, Quantitative determination of pyridine adsorbed on silver electrodes, *J. Electroanal. Chem. Interfacial Electrochem.* 1979, **105**, 409–411.

[20] C.Y. Chen, E. Burstein and S. Lundquist, Giant Raman scattering by pyridine and cyanide(−) ion adsorbed on silver, *Solid State Commun.* 1979, **32**, 63–66.

[21] J.A. Creighton, C.G. Blatchford and M.G. Albrecht, Plasma resonance enhancement of Raman scattering by pyridine adsorbed on silver or gold sol particles of size comparable to the excitation wavelength, *J. Chem. Soc., Faraday Trans. 2* 1979, **75**, 790–798.

[22] B. Pettinger, U. Wenning and H. Wetzel, Angular resolved Raman spectra from pyridine adsorbed on silver electrodes, *Chem. Phys. Lett.* 1979, **67**, 192–196.

[23] B. Pettinger, A. Tadjeddine and D.M. Kolb, Enhancement in Raman intensity by use of surface plasmons, *Chem. Phys. Lett.* 1979, **66**, 544–548.

[24] R.R. Smardzewski, R.J. Colton and J.S. Murday, Enhanced Raman scattering by pyridine physisorbed on a clean silver surface in ultrahigh vacuum, *Chem. Phys. Lett.* 1979, **68**, 53–57.

[25] J.P. Heritage, J.G. Bergman, A. Pinczuk and J.M. Worlock, Surface picosecond Raman gain spectroscopy of a cyanide monolayer on silver, *Chem. Phys. Lett.* 1979, **67**, 229–232.

[26] C.S. Allen and R.P. Van Duyne, Orientational specificity of Raman scattering from molecules adsorbed on silver electrodes, *Chem. Phys. Lett.* 1979, **63**, 455–459.

[27] R.M. Hexter, Enhanced Raman intensity of molecules adsorbed on metal surfaces. Experiments and theory, *Solid State Commun.* 1979, **32**, 55–57.

[28] J.C. Tsang, J.R. Kirtley and J.A. Bradley, Surface-enhanced Raman spectroscopy and surface plasmons, *Phys. Rev. Lett.* 1979, **43**, 772–775.

[29] G.L. Eesley and J.R. Smith, Enhanced Raman scattering on metal surfaces, *Solid State Commun.* 1979, **31**, 815–819.

[30] R.M. Hexter and M.G. Albrecht, Metal surface Raman spectroscopy: theory, *Spectrochim. Acta, Part A* 1979, **35**, 233–251.

[31] M. Moskovits, Enhanced Raman scattering by molecules adsorbed on electrodes – a theoretical model, *Solid State Commun.* 1979, **32**, 59–62.

6

SERS/SERRS, the Analytical Tool

Raman experiments are designed to detect the inelastic scattering from the analyte molecule. In a steady-state experiment, the measured Raman signal is characterized by its peak position, integrated band intensity and bandwidth [full width at half-maximum (FWHM)] discussed in Chapter 1. In addition, polarization of the exciting light and polarization analysis of the scattered light provide a powerful experimental tool that gives rise to a host of Raman techniques that come under the umbrella of polarization spectroscopy [1]. In the nonenhanced Raman experiment, the three parameters are the results of averaging, and therefore they do not change or deviate with time. Quantitative chemical analysis is routinely carried out with the help of well-characterized standards using average Raman signals [2]. The analytical application of Raman scattering is an expanding field boosted by low-cost lasers, miniaturization of Raman systems and the introduction of fiber-optic probes and smart instrumentation [3–11]. However, the straightforward application of the analytical methodology is not always possible in the case of SERS experiments. More often than not, signal will be lost, bands will shift, random peaks will appear, bandwidths will widen and intensities may change with time. The problems intensify at trace detection levels where the number of molecules of the analyte is at sub-monolayer coverage of the enhancing surface area involved. Metal nanoparticles and their aggregates that support surface plasmon resonances form the nano-SERS regime, and the SERS signal originates from the few molecules residing

Surface-Enhanced Vibrational Spectroscopy R. Aroca

in so-called 'hot spots'. The characterization of hot spots has proven elusive and is ongoing. As a result, any slight perturbation of the targeted molecules or of the hot spot itself will have drastic effects on the detected signal.

Despite the shortcomings, SERS is now accepted as an analytical technique, and it is widely used as a powerful tool for ultra-sensitive chemical analysis down to single molecule detection (SMD) [12–17]. The main hurdle for its application in quantitative analysis is the lack of a reliable and reproducible homogeneous SERS *substrate*. There has been, and continues to be, a flurry of activity to solve this major problem. The large variety of substrates tested for SERS is widening and we are almost at a point where there is a substrate for every need. The task of finding 'the universal SERS substrate' is not trivial and may not even be the right approach for implementing SERS as an analytical tool. For practical quantitative analytical applications, SERS, as a tool, must fulfil the typical requirements of an analytical technique: reproducibility of the results, linearity of the response, standards, molecular selectivity and clear methodology for sample preparation. Unfortunately, these requirements are not easily met in the case of SERS experiments. However, there is a partial solution to the analytical problem based on experimental results that entails the recognition of two SERS 'regimes', and is a separation based on the magnitude of the enhancement factor. As was defined in Chapter 3, at low enhancement factors (up to $\sim 10^6$), the good-quality signal from the statistical average SERS of an ensemble of scatterers, called 'average SERS', or as Otto *et al.* have described it, canonical SERS [18], normally produces stable and reproducible spectra with well-defined average band centers, FWHMs and relative intensities. The second regime is the nonaveraged SERS obtained from a small number of target molecules adsorbed on nanostructures with very high enhancement factors ($\sim 10^{10}$ or higher). The latter is now commonplace for the high enhancement factor SERS observed on nanostructures that support multimode plasmons, such as fractals or aggregates of particles. Here the SERS spectrum is dominated by the scattering originating from the few molecules residing in hot spots, a name reserved for the spatial location producing an ultra-high enhancement factor. The dynamics characteristic to this regime may lead to fluctuating signal intensity. These fluctuations are the result of both photo-induced and spontaneous dynamics [19], including photodesorption, which is commonly observed in the visible and even in the infrared region, where it is thermal and indiscriminate [20]. Perturbations can be separated into those that are transient, temporal and nondestructive, and those that are destructive.

The problem here is compounded by the fact that the enhancing nanostructure has its own dynamics, especially in the presence of a radiation field. Therefore, for practical applications it is convenient to separate the discussion of analytical SERS into that of average SERS, and ultrasensitive trace analysis down to SMD.

First we will review SERS substrates commonly used for average SERS applications that include, among others, metal colloids, metal island films and electrochemically roughened electrodes. It is worth mentioning a recent book (reference 21, p. 186), where the synthesis of metal nanoparticles is discussed, and the book edited by Feldheim and Foss [22], where synthesis of metal nanoparticles (colloids, nanorods and electrochemical synthesis), optical characterization and aggregation are described in detail.

6.1 AVERAGE SERS ON METAL COLLOIDS. PREPARATION AND PROPERTIES

Since Faraday's pioneering work [23], the main aim has been to manipulate the parameters to control the synthesis of colloidal metal nanoparticles. The number of methods for the preparation of colloidal nanostructures with SERS activity is numerous and continues to be a very active field of research. Therefore, this chapter has been necessarily restricted to some of the most widely used methods based on chemical reactions in solution (wet chemistry) that yield metal nanoparticles. Metal colloids can be prepared by a variety of different procedures: chemical reduction, laser ablation and photoreduction are those most frequently employed. By far the most universally used method for the preparation of metal nanoparticles in suspension for SERS is chemical reduction. Wet chemistry is usually performed by using a starting metal salt, which is reduced by a chemical agent to produce colloidal suspensions containing nanoparticles with variable sizes, depending on the method of production. Generally, the size regime relevant to SERS experiments is between 10 and 80 nm. These particles will therefore support different plasmon resonances, depending on the size, shape and dielectric constant of the metal. The size and shape parameters can be partially controlled by appropriate choice of preparation methods. The most important parameters in this regard are the nature of the metal, the reducing reagent, the temperature, the stabilizing agents and the metal ion concentration.

Creighton *et al.* [24] published in 1979 the first report describing the use of metal sols to obtain the SERS spectra of pyridine. They used silver sols obtained by the reduction of $AgNO_3$ with $Na(BH_4)$ and gold sols by the reduction of $K(AuCl_4)$ with $Na(BH_4)$. In 1980, Kerker *et al.* [25] reported SERS from citrate ion on Ag sols. Metal sols have since become one of the standard substrates for SERS and they provide one of the most reliable SERS methodologies to obtain average enhancement values. A brief review of the most common recipes to prepare silver, gold and copper colloids in water is presented below.

6.1.1 Silver Colloids

These colloids can be prepared by the following methods [26–30].

6.1.1.1 *Citrate Colloid*

A 200 mL volume of 10^{-3} M $AgNO_3$ aqueous solution is heated to boiling, then 4 mL of 1% trisodium citrate solution are added, keeping the mixture boiling for 1 h. The resulting colloid shows a turbid gray color with an absorption maximum at ca 400 nm (Figure 6.1). This colloid is stable for months, although its nanoscopic properties may change with time. The plasmon absorption of the colloid and the size distribution

Figure 6.1 Typical external appearance of silver citrate colloids commonly used in SERS experiments

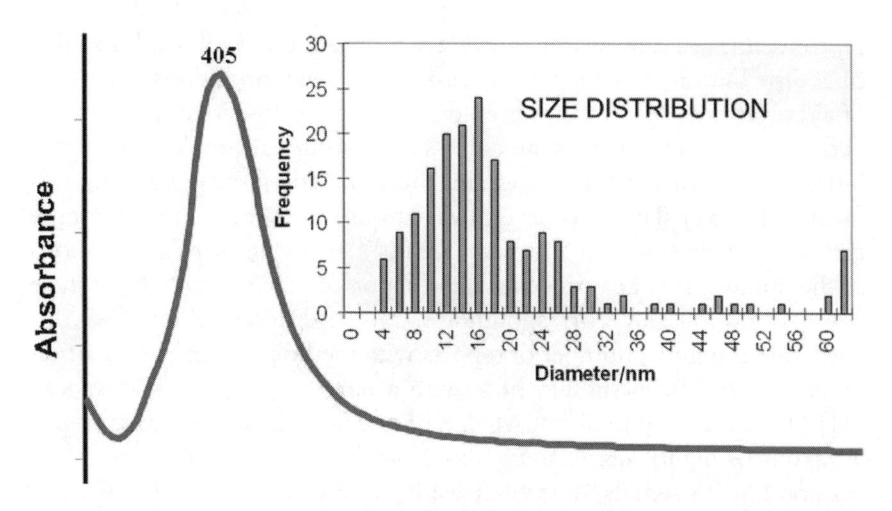

Figure 6.2 Plasmon absorption of the citrate colloids. The inset shows the particle size distribution of the cast colloids on the mica substrate

found in a detailed atomic force microscopy (AFM) study of one silver citrate solution are shown in Figure 6.2.

The AFM image recorded in air for a solution cast and dried on freshly cleaved mica is shown in Figure 6.3. The dimension of 1 μm shown is very relevant to SERS experiments, since in Raman microscopy a spatial resolution of 1 μm^2 is routinely achieved with visible lasers. The results presented in Figures 6.2 and 6.3 explain the difficulty of achieving

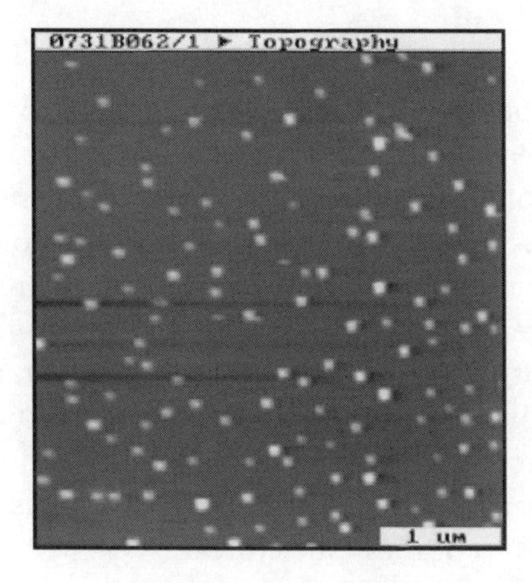

Figure 6.3 Atomic force microscopy image of citrate colloids cast on mica

reproducible SERS measurements. Each preparation of sols leads to a distinct distribution, and when using cast colloids the formations of clusters of aggregated metal particles provide the most SERS-active substrate. In addition, it is possible to tune colloidal substrates for maximum signal intensity by inducing aggregation in solution, adding salts or adding surfactants [31,32]. There are several variations of the recipe for the preparation of silver sols. For instance, Kerker's group improved variations of the initial Carey Lea procedure and reported achieving better particle size distribution and, correspondingly, sols of different colors [30]. Indeed, one can find a number of reports with methods for the preparation of silver colloids, including those with a narrow range of particle size [21] However, empirical knowledge has accumulated enough evidence in favour of highly aggregated colloids as the best substrate for obtaining good SERS signals, and what we have said for silver colloids is, in general, valid for all metal colloid used in SERS.

6.1.1.2 *Borohydride Colloid*

A 30 mL volume of 10^{-3} M $AgNO_3$ aqueous solution is added dropwise to ice-cold 2×10^{-3} M sodium borohydride solution with vigorous stirring. The mixture is kept for 1.5 h without agitation and then vigorously stirred for 10 min. The resulting colloid is stable for months and shows a yellow color with an absorption maximum at about 390 nm.

6.1.2 Gold Colloid

Gold colloid can be prepared as follows: 0.1 mL of $HAuCl_4$ solution (4 %, w/v) is added to 40 mL of triply distilled water, then 1 mL of trisodium citrate solution (1%, w/v) is added dropwise with stirring. The resulting mixture is boiled for 5 min. The gold colloid obtained shows an absorption maximum at about 525 nm and a homogeneous particle distribution (Figure 6.4).

The plasmon absorption and the optical image of a cast gold colloid dried in air are shown in Figure 6.5. The bar on the bottom-right of the optical image represents 20 μm. The cast colloids on glass and quartz form fractal-like structures and strong SERS signals are normally obtained in the branches of these structures.

Figure 6.4 External appearance of the gold citrate colloids used as SERS enhancing nanoparticles

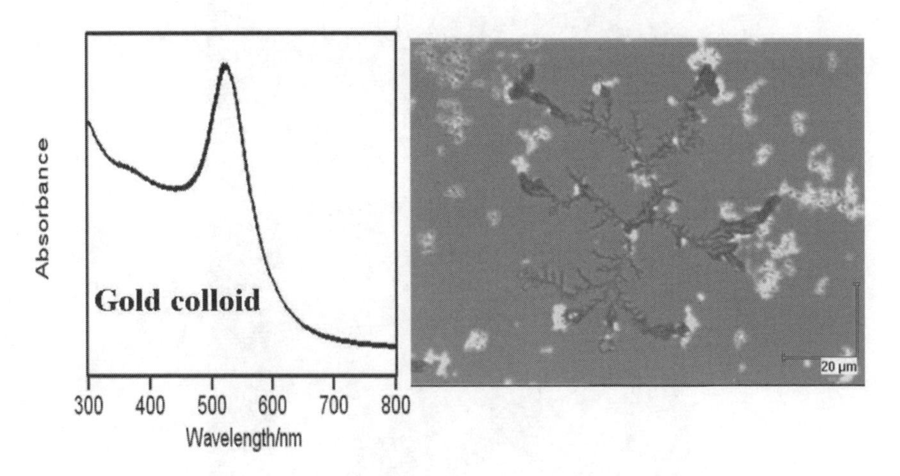

Figure 6.5 Plasmon absorption spectrum of gold colloids and the optical image of cast gold colloids on glass and dried in air. Gold colloids cast on glass and quartz form this fractal-like structure that seems to provide a strong SERS signal

6.1.3 Copper Colloid

Copper colloids can be prepared by the following procedure [27]: 5 mL of an aqueous solution of copper(II) sulfate (10^{-2} M) are added to 60 mL of trisodium citrate solution (5.6×10^{-3} M), then 30 mL of a freshly prepared solution of sodium borohydride (2×10^{-2} M) in sodium hydroxide (2×10^{-2} M) are added. The resulting colloid is yellow–brown. This colloid is aged for about 1.5 h in order to allow partial aggregation, becoming dark red (Figure 6.6). After this time, an absorption maximum at about 560 nm is observed.

Wet chemistry continues to provide an increasing number of ways to reduce gold and silver salts to form nanoparticles of different size and shape. For instance, single-crystal nanoplates with thicknesses <30 nm, characterized by hexagonal and truncated triangular shapes bounded mainly by facets, can be obtained by reduction of gold chloride with aspartic acid [33]. It was also shown that the reduction to produce gold nanoparticles can be achieved also with tyrosine, phenylalanine, lysine, and tryptophan. Triangular silver particles have been prepared by

Figure 6.6 External appearance of the aqueous copper borohydride colloids

reduction of silver ions on silver seeds with ascorbic acid in an alkaline solution of highly concentrated cetyltrimethylammonium bromide [34]. Sun and Xia [35] synthesized monodisperse samples of silver nanocubes by reducing silver nitrate with ethylene glycol in the presence of poly vinylpyrrolidone (PVP). Bimetallic nanoshells [36] and single-crystal gold nanorods and nanowires have been prepared in solution using a one-step, seedless and template-free microwave polyol method [37]. The rapid expansion of this field will provide nanoparticles of the desired size and shape for SERS applications. In our own group, we have reported syntheses of gold nanoparticles mediated by the biopolymer chitosan to form self-supporting thin films from the resultant gold–chitosan nanocomposite solutions [38] and nanowires in layer-by-layer films [39].

6.2 METAL COLLOIDS. THE BACKGROUND SERS

Since colloid preparation is an oxidation–reduction reaction involving salts and organic reducing agents (including stabilizers), a common problem is surface contaminants, mainly organic contaminants, and, in addition, what seems to be omnipresent, amorphous carbon from organic photodissociation at the metal surface. An example of each problem can make the case clear. The contamination case prompted Sanchez-Cortes and Garcia-Ramos to list a number of spurious lines that have beleaguered SERS experiments attained with metal colloids [40]. The carbon case has a long history that was recently raised again by Otto [41], who revisited the carbon contamination problem for the case of SMD. A typical example of the background SERS spectra observed on gold colloids is shown in Figure 6.7. The background spectra of the Au citrate sols in solution and the same colloids cast on glass surface are shown.

In all ambient condition experiments carried out in colloidal solutions or cast colloids, metal films, and electrochemical substrates, the target molecule at the metal nanostructure can readily interact with its surroundings, and the metal nanostructure may enhance the Raman or other signal (fluorescence) from other chemicals in the milieu. When the working conditions are those of the ultrasensitive SERS, further complications arise due to the peculiar properties of the scattering from molecules in highly localized fields. In average SERS, the most obvious interference would be the Raman scattering signal arising from the surrounding medium (water or any other solvent in use), from chemical contaminants adsorbed at the enhancing surface or contaminants that are formed at the surface as by-products of the laser radiation.

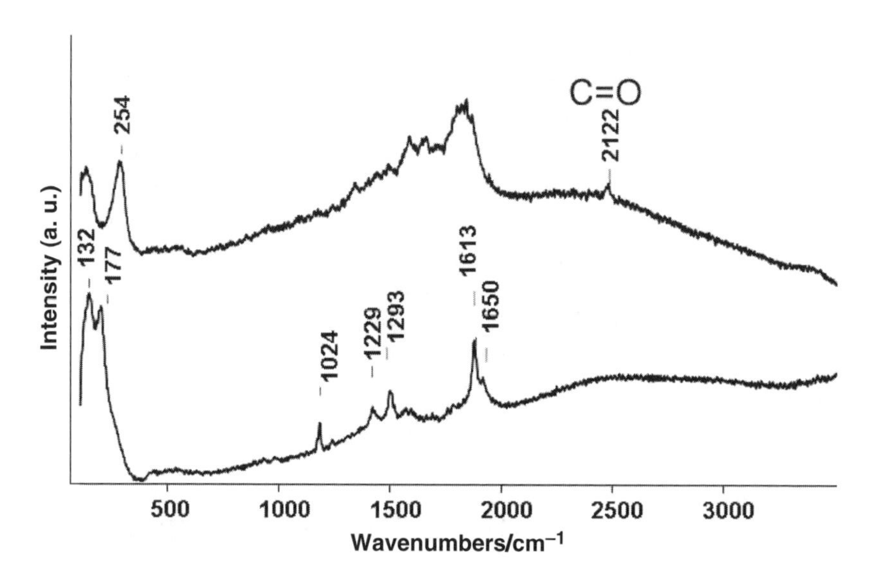

Figure 6.7 Typical Raman scattering background recorded from gold colloids in solution (bottom) and that from colloids solution cast on glass (top)

The characteristic 'cathedral peaks' detected around 1360 and 1560 cm^{-1}, which are due to graphitic or amorphous carbon, are a common occurrence in the SERS literature for a variety of substrates and detailed studies of this omnipresent SERS spectrum have therefore been carried out. It has been determined not only that disordered carbon has a very high Raman cross-section, nearly four times that of benzene [42], but also that carbon systems have a varying degree of molecular resonance throughout all the spectrum from the UV to the near-infrared region [43]. In the case of silver surfaces prepared in different ways, carbon has been always unambiguously detected [44]. A demonstration of this ubiquitous carbon signal is shown in Figure 6.8, where a single colloid cluster without an 'analyte' is examined in a series of 17 spectra taken 1 s apart with 1 mW of 514 nm excitation. The signature carbon signal can be seen but, on further examination under constant radiation, the spectrum varies. The nature of this variation has been addressed by Pettinger and Kudelski [45], who attributed the variety of bands to the modification of the graphitic-like carbon, both structurally and chemically, as it interacts with its environment. Bjerneld *et al.* [46] tracked the origin of this phenomena by looking at the photochemistry on silver nanocrystals. The experimental work also showed that the presence of oxygen could play a role in the observed spectral fluctuations [47].

In colloids and films, very common spurious bands at about 2120–2130 cm^{-1} have been attributed to CO and are especially prevalent on

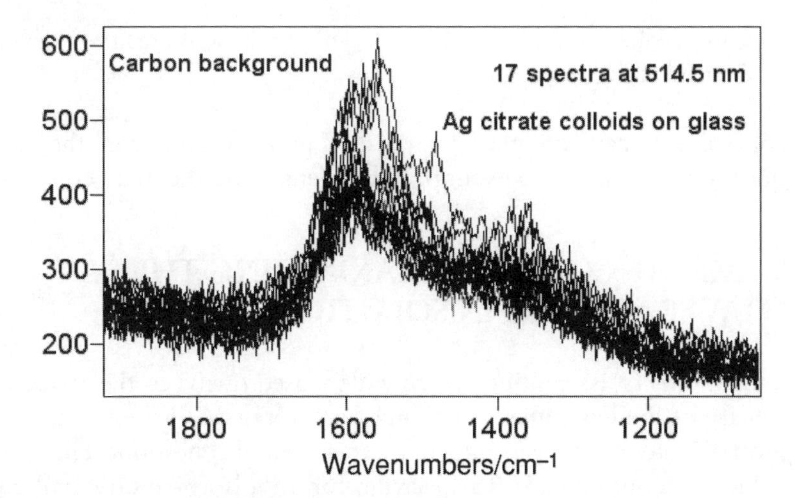

Figure 6.8 Characteristic background of silver colloids cast on glass shown by superimposing 17 spectra recorded with 514.5 nm laser excitation

gold surfaces [48]. The region below 300 cm^{-1} often contains several bands. These bands are the result of surface species complexed to the metal such as Ag–O, Ag–Cl or other molecule–metal complexes.

Laser-ablated silver colloids have been prepared [49,50], with the purpose of avoiding surface impurities introduced during the chemical preparation. The size of metal colloids formed by ablation is typically on the order of 20 nm, and the size distribution is usually broad and asymmetric. The general fabrication of silver ablated colloids involves irradiating a cleaned silver foil (99.99%, mm), immersed in doubly distilled, deionized water (specific resistance of 18 $M\Omega\,cm^{-1}$) in a quartz cell, with an Nd:YAG pulsed laser beam operating at 10 Hz with 20 ns laser pulse duration. Optimization of the ablation process can be achieved by varying several parameters, such as excitation wavelength, introducing specific ions into the aqueous media during the ablation process and focusing or defocusing the laser spot on the sample. Although silver colloids prepared by ablation are the most common, other metals such as copper have been reported. The method has been used occasionally in the literature.

In summary, there are at least three sources of interference when recording average SERS spectra in sols:

1. Spurious signals due to chemicals in solution, including graphitic carbon and CO.
2. The strong and broad water signal in the 3100–3600 cm^{-1} region that prevents the observation of vibrational frequencies above 3100 cm.

3. Low-frequency signals (150–250 cm^{-1}) whose intensity is dependent on the laser line.

For colloids cast on glass, impurities and variations in the low-frequency region with the excitation wavelength are the main concern.

6.3 METAL COLLOIDS. MAXIMIZING THE AVERAGE SERS IN SOLUTION

Metal colloids (sols) continue to be widely used owing to the attractive simplicity of the experiment allowing for 'average' SERS enhancement (up to 10^6) under fairly reproducible experimental conditions. However, to achieve an optimum SERS spectrum for an adsorbate on metal colloids, it is necessary to optimize the nanoparticle surface charge, a property of the utmost importance, as the adsorption of analyte molecules on colloidal particles is a primary prerequisite condition for obtaining strong surface enhancement of Raman signals. For instance, when colloidal particles and analyte molecules have charges of the same sign, the adsorption process can be strongly hindered, or prevented altogether. Therefore, controlling the surface charge gives the experimenter access to the kinetics of the adsorption process. On the other hand, if colloidal particles fail to exceed a minimum repulsion with one another, they will aggregate and precipitate out of solution. Clearly, particle surface charge plays an essential role in determining the stability, adsorptivity and electrokinetic properties of metal colloids, all of which are variables that govern enhancement factors in SERS/SERRS experiments. Colloidal metal particles in solution develop a net surface charge that affects the distribution of ions in the neighboring interfacial region, resulting in an increased concentration of counterions close to the surface, forming an electrical double layer (EDL) in the region of the particle–liquid interface. A detailed discussion of the finer points of EDL around colloidal particles can be found in Myers' book (reference 51, Chapter 5, p. 79). In general, the decay of the surface potential with distance is an exponential, where the decay rate slows with increase in concentration. The surface charge of colloidal nanoparticles can be monitored by means of measurement of the potential at the interface between the moving and the stationary solvent layers at their edges (i.e. the slipping plane) [52]. The slipping plane (also termed shear plane) is an imaginary surface separating the thin layer of liquid bound to the solid surface and showing elastic behavior compared with the rest of liquid, which shows normal viscous behavior.

The electric potential at the slipping plane is referred to as the zeta potential (ζ), and provides useful information about the charge carried by the nanoparticle and therefore about its stability and ability to interact with analyte molecules. Zeta potential measurements are a key factor in the preparation of colloidal dispersions for applications in paints, inks, pharmaceutical and cosmetic preparations, food products and many others. A cartoon of the metal colloid double layer is shown in Figure 6.9. In a first approximation, the *electrophoretic mobility* (the ratio of the velocity of particles to the field strength), the induced pressure difference in electro-osmosis, streaming potential and the sedimentation potential are proportional to the zeta potential. The stability of hydrophobic colloids depends on the zeta potential: when the absolute value of the zeta potential is above 50 mV, the dispersions are very stable owing to mutual electrostatic repulsion, and when the zeta potential is close to zero, the *coagulation* (formation of larger assemblies of particles) is very fast, which causes fast sedimentation. Even when the surface charge density is very high but the zeta potential is low, the colloids are unstable. Zeta potential measurements are directly related to the nature and structure of the electric double layer at the particle–liquid interface.

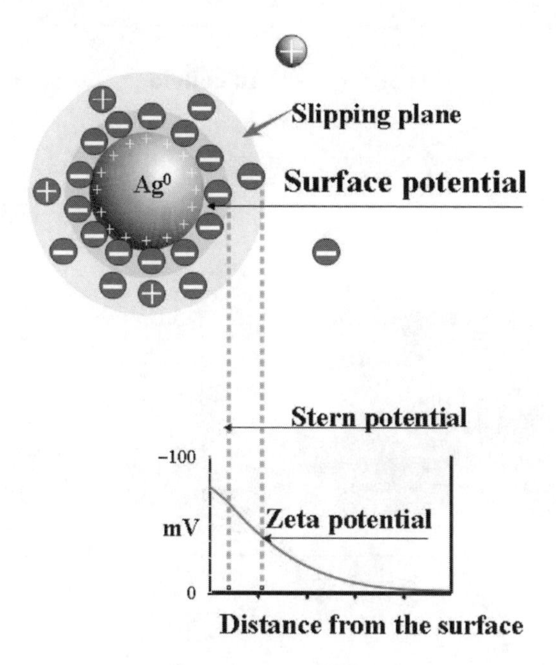

Figure 6.9 Cartoon representing the metal colloid double layer, illustrating the nomenclature for the distance dependence of the surface charge

There are reports in the literature correlating the zeta potential with the properties of sols. Lecomte *et al.* studied the zeta potential of silver colloidal solutions prepared by citrate reduction as a function of the concentration of the analyte added [53]. Faulds *et al.* studied the rate of aggregation of these colloids as a function of zeta potential by altering the same variable [54]. It is therefore important to correlate the electrostatic interactions between colloidal nanoparticles and analyte molecules and to monitor their impact on the enhancement of Raman signals in SERS experiments.

Zeta potential in metal colloids is closely related to the pH value of the solution for which it is measured. The impact of pH values and the addition of electrolytes on SERS experiments has been recognized and reported [55]. Experiments directed at establishing the direct correlation between zeta potentials and SERS enhancement factors in colloidal solutions have been carried out in our laboratory. The variation of the zeta potential as a function of pH is shown in Figure 6.10 for Au colloids prepared by reduction with citrate.

This strong variation of the zeta potential with pH would correlate with the variation of SERS intensities that would be observed when using

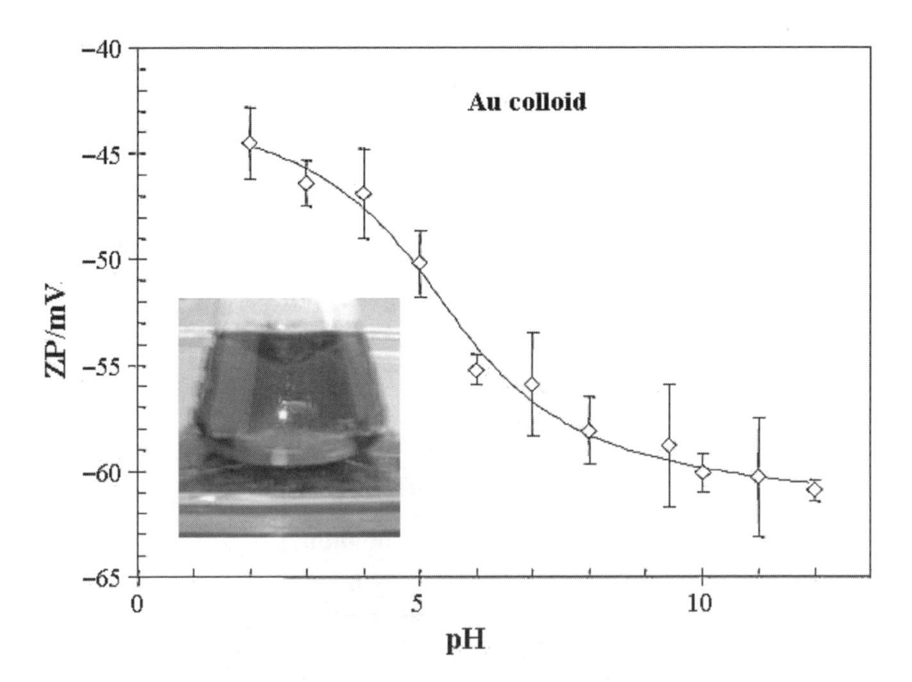

Figure 6.10 Typical experimental results for measurements of the zeta potential as a function of pH in citrate gold colloids

these solutions with analytes of differing acidic properties at variable pH. The colloidal nanoparticle solutions obtained by citrate reduction of gold shown in Figure 6.10 show negative zeta potential values (more negative than the corresponding silver citrate colloids) and high stability. The observed absorption spectra show no shifts in the wavelength corresponding to the plasmon absorption (maximum at 532 nm) of the Au nanoparticles from pH 2 to 12, but the intensity of the plasmon band is unmistakably lower at pH 2 than at pH 3, and this is probably due to the partial lost of gold particles. In addition, these gold colloids do not exhibit oxidation with pH variation, as is consistent with the high reduction potential of this metal (1.50 V). Molecules with different acid–base properties can be used to learn about the SERS dependence on the surface charge. As an example, the SERS spectra for pyridine on gold–citrate colloids as a function of pH are shown in Figure 6.11. These colloids present the most negative zeta potential values of all the most commonly used sols, from −45 mV at pH 2 down to −62 mV at pH 12. This large negative charge makes the adsorption of negative ions on the gold surface impossible, and will result in a lack of SERS signals at any pH value for negatively charged analytes. Pyridine, in contrast, does show SERS activity (Figure 6.11), owing to its lower acidity and

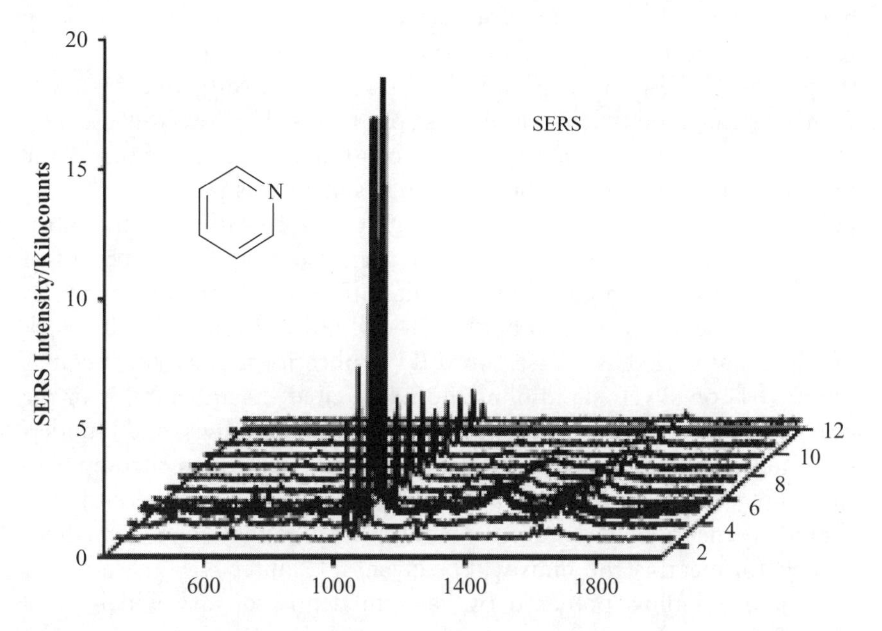

Figure 6.11 SERS intensity of pyridine is shown to decrease progressively with increase in pH, owing partly to a decrease in zeta potential values

higher basic character, but it is fairly weak in comparison with the results obtained using silver colloids, which are less negatively charged. In addition, the maximum SERS intensities for these molecules are obtained at pH 3–4, when the zeta potential values are less negative and hindrance to analyte adsorption is reduced. It can be seen in Figure 6.11 that the SERS intensity progressively decreases with increasing in pH, owing to decreased zeta potential values, and consequent increases in electrostatic repulsion.

The experimental results shown in Figure 6.11 illustrate that there is a need to tune the surface charge in the colloidal solution to obtain the maximum enhancement. In this case, the best SERS results are obtained in a very narrow region of pH values, and the intensity of the SERS spectrum at pH 3–4 is highest. However, most SERS experiments are carried out at pH 8, which is the normal pH of a colloidal solution after its synthesis.

6.4 AVERAGE SERS ON METAL ISLAND FILMS

The absorption and scattering of light by small metal particles (see Chapter 2) occur at lower frequencies than the plasma frequency of the bulk material. These lower frequencies are called surface plasmons, and are responsible for the color found in aggregated metal particles. The optical properties of these metal island films have been investigated for more than a century, and the first model was put forward by Maxwell-Garnett [56]. The complete solution for the absorption and scattering of a plane monochromatic wave by a homogeneous sphere was published by Mie [57] in 1908 (Chapter 2). There is an extensive literature on the fabrication, characterization and determination of the optical properties of metal island (aggregated) films. The literature on metal films has been collected in several excellent books [58–60]. As an example of the work on films that served as a basis for SERS applications, Yamaguchi *et al.*'s report [61] on silver island films should be cited. As an example of the pre-SERS work on gold island films, the work of Granqvist and Hunderi [62] and the excellent review by Papavassiliou [63], which encompasses the optical characterization of island films of varying shape and size, should be mentioned. An extensive discussion of the effective medium models for metal island films will be given in Chapter 7.

Thin metal films prepared by vacuum deposition are widely used as SERS substrates. Vacuum deposition has several main advantages over chemical techniques, including applicability to any substance, high purity, in some circumstances pre-selected structure, variable substrate temperature and access to the surface during deposition. The magnitude of surface enhancement from thin metal films is critically dependent on film morphology; and the factors leading to reproducible vapor-deposited thin metal films have been examined [64] and the SERS activity has been compared for some of the fabricated films [65].

The process of film formation by vacuum deposition consists of several physical stages:

1. transformation of the material to be deposited by evaporation or sublimation into the gaseous state;
2. transfer of atoms from the thermal evaporation source to the substrate;
3. deposition of these particles on the substrate;
4. rearrangement or modification in their binding on the surface of the substrate.

Film structure can be controlled by the deposition rate, substrate roughness, temperature of the substrate during deposition and mass thickness. Subsequently, the film may be annealed to activate grain growth, alter stoichiometry, introduce dopants or cause oxidation.

In physical vapor deposition, the source of the film-forming material is a solid, which needs to be vaporized so that it may be transported to the substrate. This may be accomplished by heat or by an energetic beam of electrons, photons or positive ions. The supply rate and contamination of the source are important concerns. The supply rate is important because film properties are influenced by both the deposition rate and the ratio in which the particles are supplied to the films. The possibility of contamination, however, extends far beyond the source and is also an issue in both the transportation and deposition processes.

The major concern in the transport step is the uniformity of the arrival rate over the substrate area. In a high-vacuum system, molecules travel from the source to the substrate in straight lines, and the uniformity is controlled for the most part by the geometric configuration of the system. The deposition step, on the other hand, is determined by both source and transport factors, together with the conditions at the deposition surface. There are three principal surface factors that determine the deposition behavior:

1. substrate surface condition, which includes roughness, level of contamination, degree of chemical bonding with arriving material and crystallographic parameters in the case of epitaxy;

2. reactivity of the arriving material, i.e. the probability of arriving molecules reacting with the surface and becoming incorporated into the film, also known as the sticking coefficient;
3. energy input to the surface, mainly influenced by substrate temperature, with a profound effect on both the reactivity of arriving material and on the composition and structure of the film.

These three factors work together to determine the structure and composition of the deposited film. This means that for the formation of films with reproducible properties, it is necessary that these parameters be constant and measurable. As a result, monitoring is important at all steps in the thin-film process. The details of the experimental techniques most commonly used for thin-film fabrication can be found in Smith's book [66]. For routine SERS applications, metal thin-film deposition may be performed using a Balzers high-vacuum system. The substrate is introduced through a load lock chamber to allow the main process chamber to remain under vacuum, thereby reducing contamination. For most experimental purposes, the substrates used are transparent and pre-cleaned borosilicate slides (Baxter, Cat. No. M6145), cleaned by rubbing them with absolute ethanol and subsequent drying under a continuous flow of dry nitrogen gas. Once the substrate is in the process chamber, it is heated and controlled at a fixed deposition temperature. The working pressure is nominally 10^{-6} Torr. The metal materials are thermally evaporated from cupped tungsten boats using a Balzers BSV 080 glow evaporation control unit. The evaporation rate is allowed to stabilize before the shutter is opened. The mass thickness of the thin films and deposition rate are monitored with an XTC Inficon quartz crystal oscillator.

On borosilicate slides preheated to 200°C, silver is deposited at a rate of 0.05 nm s^{-1} to a total mass thickness of 6 nm for silver island films for SERS substrates. Similarly, silver may also be evaporated on KBr discs to a mass thickness of 6 nm. The bulk density of silver employed is 10.5 g cm^{-3}, the tooling factor 105 % and the *Z*-ratio 0.529. For surface-enhanced infrared experiments, silver and tin may be evaporated on ZnS substrates. In such instances both silver and tin are deposited on preheated ZnS substrates, 200°C for Ag and 80°C for Sn, at a rate of 0.05 nm s^{-1} to a total mass thicknesses of 10 and 18 nm, respectively. The aforementioned evaporation parameters for silver are used. The bulk density of tin employed is 7.30 g cm^{-3}, the tooling factor 105 % and the *Z*-ratio 0.724 [67,68].

Mixed films of silver and gold with a total mass thicknesses of 10 nm are prepared in the same fashion. These mixed metal substrates are prepared by two separate evaporation procedures. First, 5 nm Ag films

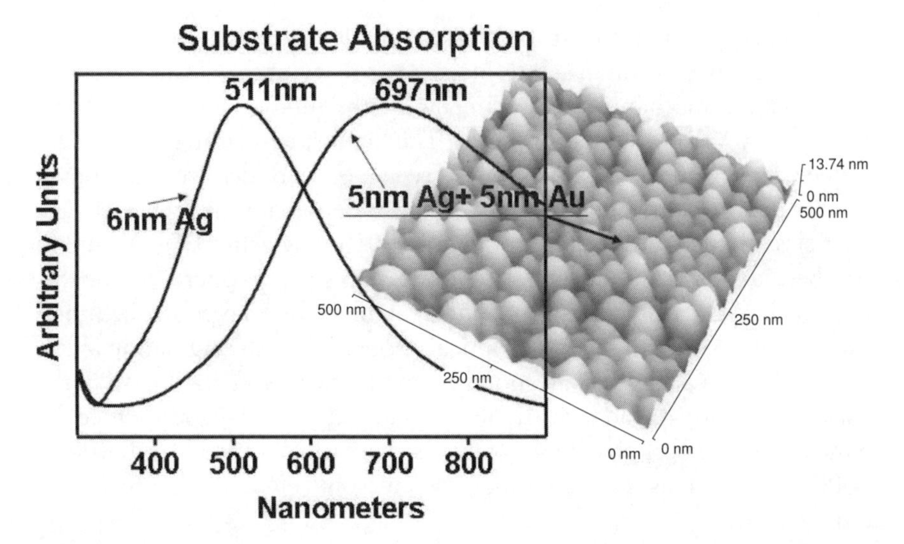

Figure 6.12 Plasmon absorption of a silver island film and that of mixed Ag/Au film. The AFM image is that of the mixed silver/gold film

are deposited on Corning 7059 glass slides, and, after allowing a sufficient cooling time (1–2 h), second layer films of 5 nm Au are then deposited on top. All depositions are carried out at evaporation rates of ca 0.5 Å s^{-1}. The spectroscopy, analysis and morphological characterization of all mixed Ag–Au substrates is accomplished using three complementary techniques: UV–visible absorption spectroscopy, X-ray photoelectron spectroscopy (XPS) and atomic force microscopy (AFM). An example of a mixed film is shown in Figure 6.12. Binary films is one of the many families of substrates that are being fabricated and tested for SERS applications [69,70].

Metal island films have become one of the most successful substrates for SERS applications and have been used from the very beginning of experimental SERS. The initial work on metal island films, mainly silver, can be found in Kerker's selected papers [71]. SERS-active substrates comprised of vapor-deposited Ag films have exhibited better stability over time, and the substrate performance has been thoroughly studied as a function of geometry, deposition rate and deposition temperature [64,72,73]. The efforts to produce films of different morphologies for SERS have produced a variety of metal film substrates. For instance, large-area arrays of metal needles have been grown and tested as substrates for SERS. These arrays are made by utilizing a fabricated base layer as a foundation for needles having a regular, readily controlled pattern of growth. Needles ca 50–75 nm in diameter with aspect ratios of 5:1 or greater are readily achieved [74].

The idea of tuning the absorption of metal island films for maximum surface-enhanced Raman scattering is a corollary of the electromagnetic enhancement mechanism. Experimentally, the tuning was achieved and reported by Van Duyne's group [75]. The 'nanosphere lithography' technique was developed in which polystyrene nanoparticles are spin-coated on to substrates and act as masks for vapor deposition of Ag metal. The particles are then removed from the surface, leaving behind Ag metal that had been deposited in the interstices of the polystyrene spheres. Using this approach, Ag feature size, shape and spacing have been controlled, permitting detailed analysis of the optical properties of the Ag nanoparticle arrays. The wavelength corresponding to the extinction maximum of the surface plasmon resonance of silver nanoparticle arrays fabricated by nanosphere lithography (NSL) can be systematically tuned from 400 to 6000 nm. The plasmon resonances taken from reference 75 are shown in Figure 6.13.

In Figure 6.13, UV–visible absorption spectra of Ag nanoparticle arrays on mica substrates are shown, illustrating the controlled variation of the plasmon (A to H) achieved with this technique. The reported spectra were raw, unfiltered data and the oscillatory signal superimposed on the plasmon spectrum seen in the data is due to interference of the probe beam between the front and back faces of the mica.

Cold-deposited silver films are disordered and porous and are generally less well characterized than the corresponding metal island films fabricated at room temperature or on heated substrates [76,77]. In an early report, Albano *et al.* [78] characterized the adsorption of pyridine on cold-deposited Ag films annealed at temperatures ranging from 58 to 330 K using ultraviolet photon spectroscopy (UPS), work function change and thermal desorption measurements. Pyridine-induced work function changes were employed to follow the surface diffusion of pyridine molecules into the pores of these Ag films. The data collected on the structure of the cold-deposited Ag films and the adsorption behavior of pyridine on these films, and also a survey of previously published SERS data, indicates that the SERS-active sites of cold-deposited Ag films are within the pores. The latter is a very important conclusion and it brings attention to the role of voids and cavities in the mechanism of SERS. Otto's group [76,77] extended the studies to physically adsorbed molecules on cold-deposited silver films in an effort to separate the different contributions to the observed SERS signal. The average SERS enhancements observed in all these experiments are between two and seven orders of magnitude.

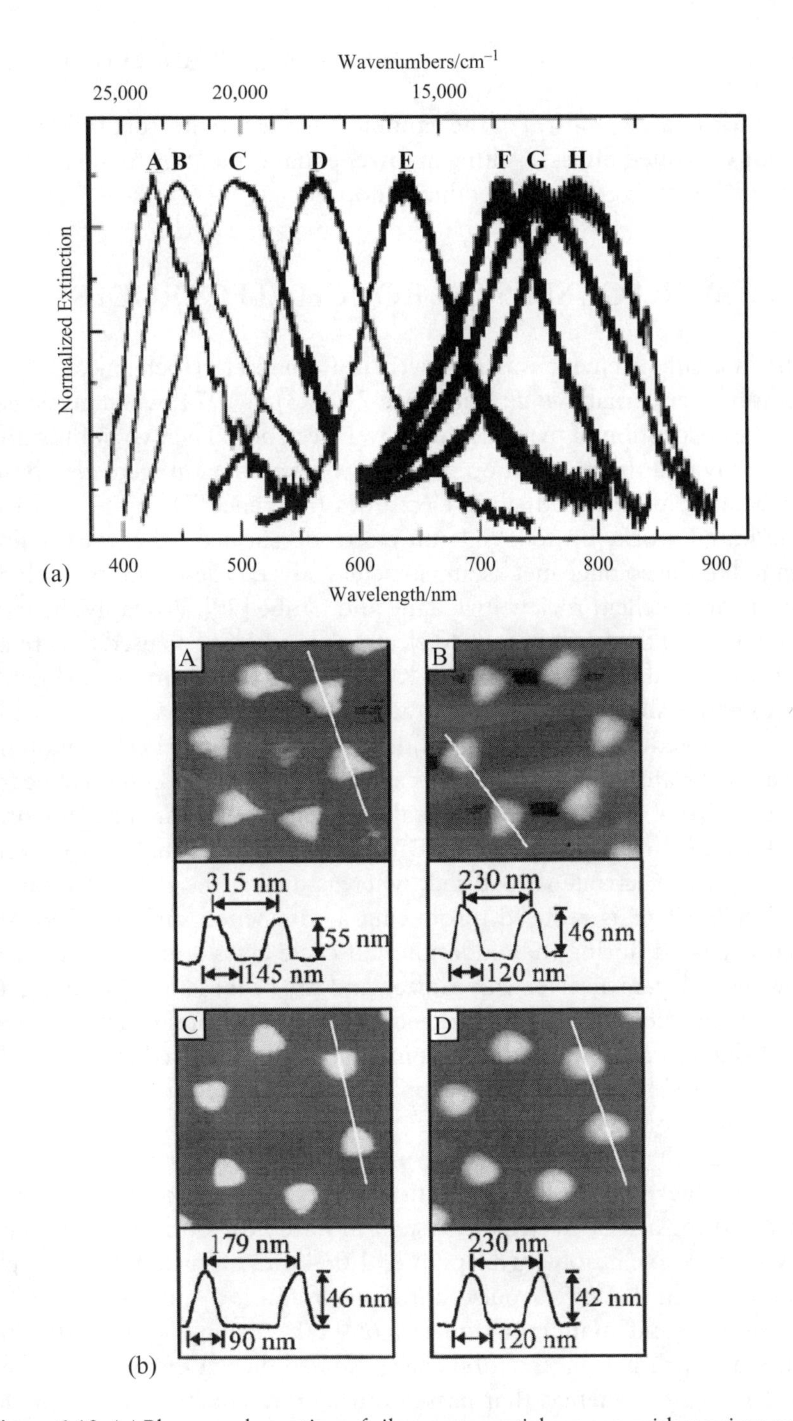

Figure 6.13 (**a**) Plasmon absorption of silver nanoparticle arrays, with maxima varying from 426 to 782 nm, and (**b**) AFM images and line scans of representative Ag nanoparticle arrays on mica substrates. Reproduced with permission from T.R. Jensen, M.D. Malinsky, C.L. Haynes and R.P. van Duyne, *J. Phys. Chem. B*, **104**, 10549–10556. Copyright 2000 American Chemical Society

Unfortunately, control of the nanometer-scale architecture is limited in vapor-deposited films, resulting in lower enhancement factors than those attained with aggregated metallic nanoparticles.

6.5 AVERAGE SERS ON ROUGH ELECTRODES

The first enhancement was observed in Raman electrochemical experiment by Fleischmann *et al.* (reference 71, p. 3) in 1974, when investigating the adsorption of pyridine on a silver electrode. The two publications that provided the recognition of the new phenomenon were also based on data from anodized silver electrodes (reference 71, pp. 7 and 27). The initial work (up to 1984) on electrode surfaces, that was mainly limited to the coinage metals, in particular silver, is described with all details in an excellent review by Chang and Laube [79]. Probably the most common and thoroughly studied electrochemical system used to activate SERS electrode substrates is the KCl electrolyte system. To achieve a SERS-active silver electrode substrate, a silver electrode is roughened by oxidation–reduction cycles (ORC) in an electrochemical cell containing an appropriate solution of a silver salt. Usually, the potential is linearly ramped from the most negative value towards the switching potential back to the most negative value. During the oxidation half of the cycle, silver at the electrode is oxidized, whereas during the reduction half of the cycle, silver is reduced, producing a roughened surface. The total charge passed during the oxidation half cycle gives a direct measure of how many layers of silver are oxidized and can be measured by taking the path integral of current with respect to time. The total charge through the reduction cycle measures how many Ag^+ ions are reduced from the adherent AgCl layer. The surface protrusions at the electrode surface are of the order of 25–500 nm in size.

Induced changes in the surface morphology of silver electrodes may also be achieved by laser illumination effects. Changes of this type were observed for a KCl electrolyte system in nonlinear optical experiments in which an excitation wavelength of 1.064 μm, an incident energy per pulse fixed at 1.4 mJ, a pulse duration of 15 ns, a pulse rate fixed at 20 pulse s^{-1} and an illumination area of 0.003 cm^2on the electrode [80]. With laser irradiation, the total charge passed during the oxidation half cycle increased whereas that passed during the reduction half cycle decreased. This is due to the fact that the AgCl layer is partly photoreduced by the laser illumination and, as a result, more silver was oxidized and fewer Ag^+ ions were electrochemically reduced [79].

Numerous other electrolytes have been tested for the fabrication of electrochemically active SERS substrates, including K_2SO_4 (which is of particular interest in the investigation of CN^- species), pure halides, sulfates and silver salts. Since the SERS signals of the anions are dependant on the solubility and adherence strength of the silver salt formed during the oxidation half cycle, then only those salts formed which are insoluble and adhere to the electrode produce SERS-active substrates. Silver salts that give an appreciable SERS signal are AgCl, AgBr, AgI, AgCN, AgSCN, AgOCN, AgN_3, Ag_3PO_4, Ag_3PO_3 and Ag_2CO_3. If the salt formed during the oxidation half cycle is slightly soluble and/or slightly adherent to the surface, such as Ag_2SO_4, $AgNO_2$ and $AgC_3H_5O_2$, the SERS signal is generally less intense and observable only within the faradaic region of the reduction cycle, and not after complete reduction. On the other hand, if the salt is highly soluble and/or only weakly adherent to the surface, such as $AgNO_3$, AgF and $AgClO_4$, then SERS of the anions is usually not observed during any part of the ORC.

Spectroelectrochemical SERS was later extended to other metal electrodes, such as platinum, palladium and iridium [81–83]. Two distinct approaches employing electrochemical methods have been attempted. Tian and co-workers [84,85] explored strategies aimed at generating SERS from transition metals by suitable surface roughening. Wasileski *et al.* [86] achieved successful SERS enhancements via electrodeposition of transition metal overlayers on SERS-active substrates via a constant-current deposition method. In this method, a cathodic current density of 40–150 $\mu A\,cm^{-2}$ (at ca −1.0 V vs SCE) for 2–3 min from dilute solutions (millimolar) of salts of the metal was used to obtain thick films of transition metals, whereas smaller currents were used to obtain the thinner films. Tian and Ren [87] gave a detailed account of surface roughening procedures for electrodes of different metals that result in good-quality SERS-active electrode surfaces made from Pt, Ni, Co, Fe, Pd, Rh, Ru and their alloys that are not traditional SERS-active substrates. These new electrochemically SERS-active substrates may find applications in co-adsorption, electrocatalysis, corrosion and fuel cell studies.

SERS and SERRS [88] electrochemistry continues to be a very active field of research given its unique ability to control and manipulate the surface charge (surface potential) [89], permitting some degree of control of adsorption and desorption [90], charge transfer phenomena [91], and molecular orientation at the surface [92,93]. Further extension of the spectroelectrochemistry into other spectral regions and applications could be part of the new SERS renaissance, as Tian and Ren [87] put it: “For three noble metals (Au, Ag, and Cu), the excitation lines have only covered the visible spectrum up to the near-infrared region, from 450 nm to 1064 nm. There are inherent difficulties in using ultraviolet excitation that could explain the absence of these data in the SERS literature. Electromagnetic enhancements are rather small in the ultraviolet region where damping is generally large due to interband transitions. Since the optical properties of transition metals are different from those of the noble metals, it is worth testing these new systems. Very recently we reported the first UV–SERS spectra of molecules adsorbed onto rough rhodium (Rh) and ruthenium (Ru) metal surfaces [94]. Hopefully, this new trend will be facilitated by the fast development of commercial low cost UV–laser and UV–Raman systems. UV–SERS would be further applied in fields that include electrochemistry, biomedicine and catalysis as well as theoretical investigation counting SERS itself.

6.6 ULTRASENSITIVE SERS ANALYSIS AND SINGLE MOLECULE DETECTION

Approaching single molecule detection using SERS was a matter of time after 1995 when Kneipp *et al.* [95] investigated the detection limits in concentration of Rhodamine 6G and Crystal Violet dye molecules in colloidal Ag solutions activated by NaCl ions. The authors reported enhanced Raman scattering spectra from less than 100 molecules with a reasonable signal-to-noise ratio. They predicted that SERS could be used to achieve detection limits comparable to those in fluorescence spectroscopy with vibrational structural specificity. With the recent advances in ultrasensitive instrumentation [3], where a microscope is used for laser excitation and collection of the signal, two groups [14,96] reported the first spectra of the single molecule in 1997. There is an important difference between these two independent results. In one case, the probe is a single Rhodamine 6G molecule adsorbed on the selected silver nanostructures using a laser line in resonance with the electronic absorption of the molecule [14]. Therefore, the observed inelastic scattering corresponds to SERRS, double R for double resonance with the particle plasmon and the molecular absorption. In the report by Kneipp *et al.* [96], the enhanced Raman scattering spectra of a single Crystal Violet molecule in colloidal silver solution was obtained using nonresonant near-IR excitation, i.e. using a laser line with a frequency outside the electronic transition of the molecule. The SMD showed well-defined fingerprints of its vibrational Raman spectrum. In both cases, Raman enhancement factors were estimated to be in the 10^{14}–10^{15} range, which are several orders of magnitude larger than the ensemble-averaged values derived from conventional measurements. The net result of these findings was that they established vibrational Raman spectroscopy as a competitive ultrasensitive technique to single-molecule fluorescence, with all the advantages of the high chemical information content of vibrational spectroscopy over fluorescence [97].

Before 1997, the SMD work had been mainly done using fluorescence signals thanks to the large cross-sections ($\sim 10^{-17}$ cm^2 molecule^{-1}) of the emission process [98]. A summary of the first 10 years (1989–99) of single-molecule spectroscopy (SMS) can be found in the review by Tamarat *et al.* [99]. There one can find a list of the main aromatic molecules used in single-molecule experiments at cryogenic temperatures. Soon after there were reviews by Moerner of 'A dozen years of single-molecule spectroscopy in physics, chemistry, and biophysics' [100] followed by 'Thirteen years of single-molecule spectroscopy in physical chemistry and biophysics' [101], and SMS is a growing new branch of spectroscopy.

Single-molecule SERS/SERRS brought about a renaissance of SERS activity with renewed vitality and a focus on the fabrication and properties of nanostructures that can provide the very large enhancement factor needed for SMD. In the same way that pyridine was the 'SERS molecule' and a great deal of the fundamental work was carried out using pyridine, Rhodamine 6G has been the 'SMD molecule', and there is a bulk of work done on single-molecule SERRS of Rhodamine 6G. Detailed studies of SMD of Rhodamine 6G have been carried out by several groups [102–106]. However, Michaels *et al.* [102], using atomic force microscopy (AFM) measurements, showed that the Ag nanoparticles that yield SMD of Rhodamine 6G (R6G) are all compact aggregates consisting of a minimum of two individual particles. SMD was quickly extended to other molecular systems. Xu *et al.* [107] achieved the detection of single hemoglobin protein molecule attached to isolated and immobilized Ag nanoparticles using SERS. They also arrived at the conclusion that SMD SERS was possible only for molecules situated between Ag nanoparticles. Kneipp *et al.* [108] reviewed SMD using SERS, with nonresonant near-IR excitation, again emphasizing the conclusion that target molecules are attached to colloidal Ag and Au nanoclusters. SERRS from single myoglobin molecules was reported [109], and the vibrational spectrum of single horseradish peroxidase molecule was detected by measuring SERS from isolated and immobilized protein–nanoparticle aggregates [110]. SERS of single-stranded DNA [111], SERRS spectra of Fe–protoporphyrin IX, adsorbed on Ag colloidal nanoparticles immobilized on a polymer-coated glass slide [112], and single-molecule SERRS of the green fluorescent protein [113], have been reported. SERRS spectra of various rhodamine dyes, of pyronine G and thiopyronine adsorbed on silver clusters with SMD were obtained with a high-resolution confocal laser microscope [114].

Although aggregated colloids have been the main source for SMD, there is a growing effort to fabricate nanostructures that can support surface plasmon resonances; these are multimode resonances of the strongly interacting metal nanoparticles. Finding the appropriate strategy that one might use to construct nanostructures that predictably and repeatedly produce the giant enhancement factor is the 'holy grail' of SMD. Etchegoin *et al.* [115] highlighted the fact that sharp plasmon resonances can be sustained inside clusters of nanoparticles, such as aggregated colloids. When approaching SMD, starting at submonolayer coverage of the metallic nanostructure, there is a trend that Pettinger [48] summarized as follows: "the rule of thumb is that the higher the surface enhancement of Raman cross-sections, the more localized are the zones of high Raman activity". These highly localized nanoscale interstitial sites in nanostructures or rough fractal surfaces are the centers of Raman enhancement, and are also the origin of hot spots. Hot spots are believed to be small subwavelength regions sustaining optically active surface plasmon resonances.

Aligned Ag nanowires have been fabricated, where "the observations can be adequately described by the classical electromagnetic (EM) response of the strongly interacting metal nanowires to the optical fields when surface plasmon resonances are induced" [116]. Au core–Ag shell bimetallic nanoparticles have been fabricated by a seed-mediated technique, and a claim was made that small particles (less than 50 nm) were found to be highly efficient for SERS (of Crystal Violet), and permit SMD of the selected dye molecule [117]. Tip-enhanced Raman scattering (TERS) [118] is a unique approach in which a metal tip is used as the nanometric enhancing unit. The metal tip is expected to provide the electromagnetic enhancement.

In our laboratory, vacuum-evaporated silver island films, having a morphology consisting of aggregated nanoparticles, as shown in the AFM image in Figure 6.12, were used. SMD is achieved by spatially resolved SERRS microscopy of a single Langmuir–Blodgett (LB) monomolecular layer containing dye molecules dispersed in a fatty acid, i.e. a two-dimensional host matrix fabricated using the LB technique, [119]. There are several advantages of working with LB films for applications in trace analysis and SMD using SERS/SERRS. The LB technique allows for the restriction of probed molecules (analyte) to a one molecule thick surface coating on metal clusters. Also, probed films can be diluted using mixed monolayers containing well-defined amounts of analyte. As a matrix for the target molecule, one can use a fatty acid such as arachidic acid (AA) ($C_{19}H_{39}COOH$). Fatty acids are amphiphiles (one end is hydrophilic and the other is hydrophobic) which form strong monolayers, and are consistently transferable with unit transfer ratios. The area per molecule occupied by arachidic acid on the surface is well known (25 $Å^2$), its hydrocarbon chain is chemically inert and it is a very poor scatterer in the region below the C–H stretching vibrations (below 2800 cm^{-1}). It provides an ideal matrix with a suitable window for vibrational studies using LB–SERS/SERRS. The disadvantage is that the technique is applicable only to molecules that can be made into transferable monomolecular layers, severely reducing the scope of target molecules. Using the LB technique for monolayer coating and metal island films as substrate, we have reported SMD for a series of perylenetetracarboxylic diimide derivatives. Notably, resonance Raman spectra of these molecules excited in the ultraviolet and visible regions are found to give overtone and combination progressions of ring stretching vibrations with high relative intensity. SERRS spectra of concentrated mixed LB monolayers reveal that fundamentals and overtone and combination progressions are enhanced with similar enhancement factors, and the progressions in SERRS spectra are observed with strong relative intensity. Using the LB technique and the spatial resolution of Raman microscopy, SMD for three different PTCD derivatives was reported [120]. In each case it was found that single-molecule SERRS allows the observation of the first overtones and combinations from three characteristic fundamental stretching ring vibrations when excited within the envelope of the visible absorption.

To facilitate SMD, after the breakdown of ensemble averaging, spatial mapping of SERRS intensities is the best recording technique in the study of mixed dye–fatty acid LB monolayers deposited on aggregated Ag island films. By variation of the doping of the analyte in the films, the effects of dye concentration on SERRS spectra are monitored down to the single molecule level. As an example, 2D mapping SERRS experiments of *n*-pentyl-5-salicylimidoperylene (salPTCD) dispersed in monolayers of arachidic acid on Ag nanostructured films are presented here. The samples are mixed monolayers of these two materials prepared at the air–water interface in a Lauda Langmuir film balance, and transferred using Z-deposition on to glass or 6 nm (mass thickness) Ag films on glass in varying concentration ratios, starting with a 10:1 AA:salPTCD molecular ratio down to the single molecule level, with one salPTCD molecule per μm^2 of surface area. All Raman scattering experiments were conducted with a Renishaw InVia system, using laser excitation at 514.5 nm (argon ion) and laser powers below 1 mW at the sample. All measurements were made in a backscattering geometry, using a 50× microscope objective with a numerical aperture value of 0.75, providing a spatial resolution for the scattering areas of ca 1 μm^2.

The molecular structure, the absorption spectrum of a 10^{-5} M solution and the plasmon resonance of the silver film are shown in Figure 6.14. It can be seen that the double resonance is achieved with any laser line in the 450–550 nm spectral region. The SERRS mapping was obtained using the 514.5 nm laser line. For all images shown here, the intensity of the strong fundamental vibrational mode of salPTCD observed at 1300 cm^{-1} has been mapped as a function of $x - y$ area components, and each pixel corresponds to a single recorded spectrum. All the SERRS mapping experiments were carried out using the rastering of a computer-controlled three-axis encoded (XYZ) motorized stage, with a step of 1 μm, where a single 1 s accumulation was recorded at each spot. This strategy serves to facilitate the fast and easy collection of large amounts of data, while also minimizing sample laser exposure times, and thus sample

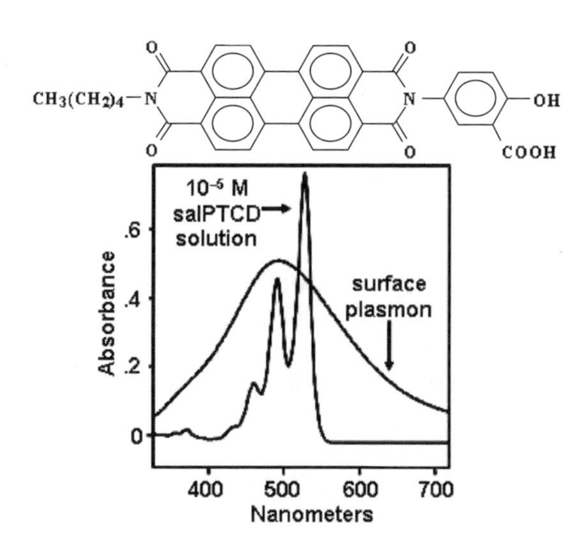

Figure 6.14 Plasmon absorption of Ag island film on glass and electronic absorption spectrum of salPTCD solution showing the double resonance that can be achieved with excitation at 514.5 nm

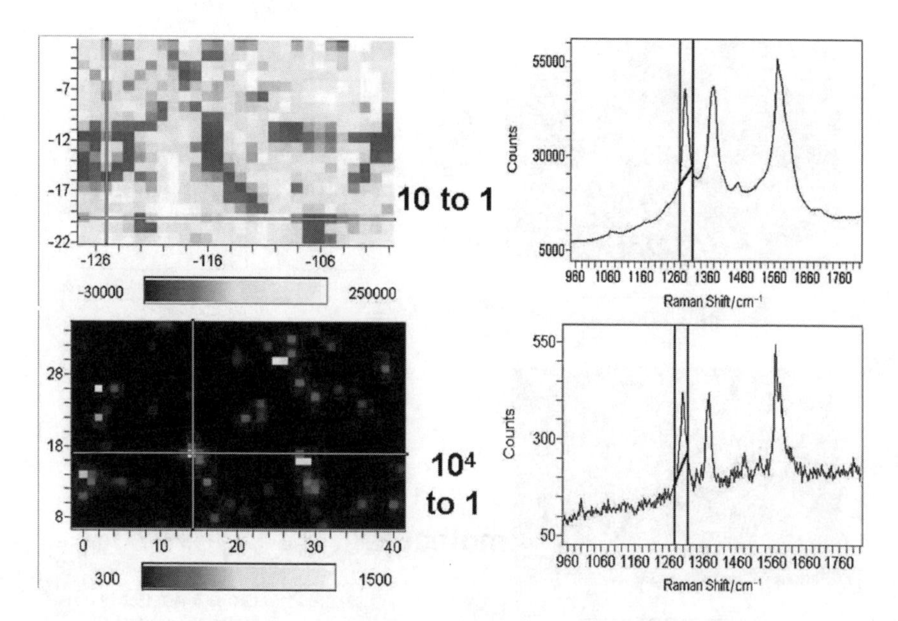

Figure 6.15 SERRS mapping and the spectra illustrate the results obtained for two concentrations of the target molecule in a monolayer of arachidic acid. The vibrational mode used for mapping is clearly shown in between parallel lines

photodegradation. At the top of Figure 6.15, the vibrational band used for mapping and the resulting map are shown for a 10:1 AA:salPTCD LB monolayer on an Ag nanoparticle film, with a spatial resolution of 1 μm^2. Essentially, strong average SERRS signals (light areas) are detected at all spots on this sample, but there are domains present in the monolayer where slight signal differences are observed, and the origin of these areas can be attributed to the aggregation of PTCD dye molecules that is known to be possible at these concentration levels in mixed LB films. Assuming that every one of the illuminated PTCD molecules contributes to the average SERRS intensity, the Raman vibrational spectrum shown at the top is the results of about 1 amol of scatterers.

The spectrum of 1 amol is the most concentrated sample in the study (apart from the neat monolayer) and is characterized by a high signal-to-noise ratio, steady-state bandwidth and frequency. The map and the spectrum at the bottom of Figure 6.15 illustrate the results obtained for a regime with the total number of scatterers in the illuminated area is ca 400 or about 1 zmol. Notably, apart from the much lower signal-to-noise ratio, the FWHM visually decreases and narrower bandwidths

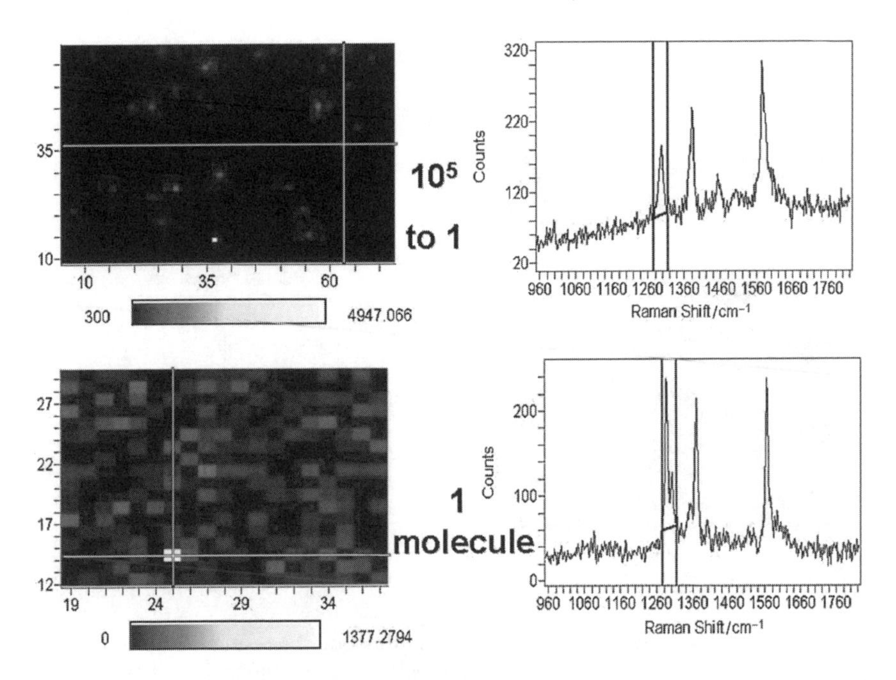

Figure 6.16 SERRS mapping measurements recorded for monolayers with spatial resolution of about 1 μm square. The spectrum of the single molecule from the monolayer is for an average of one molecule within the spatial resolution of the Raman microscope. The number of spectra collected for each map was only between 1000 and 2000 spectra

are registered, and, at the same time, changes in the relative intensity of vibrations are observed.

Below the zeptomole concentration of the target molecules the maps are similar, in the sense that spectra are obtained from a few spots within the scanned surface area as can be seen at the top in Figure 6.16. At concentration ratios of 10^4:1 and 10^5:1 AA:salPTCD, the breakdown of ensemble averaging is observed for the spectra recorded from these samples, as can be seen by comparison with the map shown in Figure 6.15 for the 10:1 sample. The result can be interpreted as an indication that, at these concentration levels, average surface enhancement is insufficient for molecular detection, and all signals are detected from local electromagnetic hot spots. From the maps presented in Figure 6.16, there are only, on average, 40 (top) and one (bottom) dye molecules in 1 $\mu m^2$of surface area, respectively. Out of a possible 4×10^6 molecular sites in a single Raman scattering area (one pixel), for each of these films, it is clear that, at these ultra-low concentrations, the rarity of coincidence between electromagnetic hot spots on Ag island films, and single isolated monomers in LB monolayers, is a low-probability event, and correspondingly there is a large number of 'empty' pixels in the map of the single molecule. It appears that even with as many as 100 molecules within the relatively large scattering area of the Raman microscope (1 μm^2), we are in fact generally detecting spectra from only a single molecule. Also, considering that the average single-molecule mapping experiment shows that less than ~0.5 % of spots are hot, it can be concluded that these hot spots are fairly highly localized on the nanoscale, and correspond to relatively few molecular sites. For mapping measurements recorded for monolayers below these concentrations, the percentage of total spots that are 'hot' tends to show significant fluctuations from map to map, and this is a result of the fact that relatively few spectra (1000–2000) are collected for each map.

In summary, the LB–SERRS results reveal significant variations, from spot to spot, in the bandwidths, frequencies and relative intensities of the three fundamental bands shown. This is typical of the unique behavior that is commonly observed as SMD limits are approached using SERRS, and reveal the breakdown of ensemble averaging, and also the wide variety of different local environments found at the heart of electromagnetic hot spots on Ag island films.

6.7 UNIQUENESS OF ULTRASENTIVE CHEMICAL ANALYSIS. THE MOVING TARGET

The accumulated experimental data seem to indicate that the very origin of ultrasensitive SERS/SERRS is to be found in aggregated nanostructures rather than in a single nanoparticle (see Chapter 3). Correspondingly, the optical properties and the dynamics of these nanostructures percolate through the collected Raman spectra of the adsorbate on reaching submonolayer coverage and the main signature of the observed spectrum is that of *fluctuation*. In fact, work trying to outline the differences between ensemble averaged spectroscopic signals and single molecule events is part of the SERS literature [121]. The frequencies, bandwidths and intensities of vibrational modes fluctuate. The entire spectrum may fluctuate and this fluctuation unfortunately, has been equated with blinking, a physical phenomenon observed in single-molecule fluorescence [122]. In fact, Jacobson and Rowlen [123] have correctly pointed out that "the term blinking is used within the SERS community to describe temporal fluctuations in the Raman spectral signal." Recent experimental results [124] show that photo-induced fluctuations seems to be site dependent, in correspondence with variation of the local field enhancement from cluster to cluster. Fluctuations of the SERS signal may be observed at submonolayer coverage of the adsorbed molecules on the hot spots, confirming that the enhancement is predominantly from these nanostructures carrying few molecules. Naturally, the SERS spectrum of the single molecule may show significant fluctuations. However, observation of Raman spectral fluctuation is not sufficient to be used as the sole signature of SMD. Fluctuations are the signature for ultrasensitive SERS, i.e. the dynamics of the adsorbate and that of the site will be reflected, and can be extracted from the recorded Raman scattering. Dynamics that are characteristic of this regime include fluctuating signal intensity, broadening, changing bandwidths, random peak appearances and complete disappearance of the signal. These fluctuations seem to be the result of both photo-induced and spontaneous dynamics [124], which can include photodesorption, which is commonly observed with visible and even infrared illumination, and is thermal and indiscriminate [20]. Recall that the conditions for SERS can also generate a host of other phenomena. These photo processes can lead to situations where the SERS signal is not in a steady state and hence not stable. This, of course, is excellent news for future developments in dynamic studies; but bad news for the future of ultrasensitive SERS in quantitative trace analysis. For instance, it has been shown that fluctuations in the shape of SERRS intensities at the SMD level can be correlated with the motion of the adsorbed molecule on the surface and could serve as a sensitive local probe for the dynamics of adsorbed molecules under ambient conditions [19]. It is safe to say that average SERS is an analytical tool for quantitative chemical analysis, whereas ultrasensitive SERS provides the basis for trace detection, including SMD, and is a probe for the dynamics of the adsorbate and its environment.

Ultrasensitive SERS is mainly realized with a selective group of nanostructures, fabricated from the most common enhancing materials, Ag and Au. Apart from the work carried out under ultra-high vacuum conditions, all enhancing structures used have their own associated backgrounds, owing to methods used in their preparation [125]. The most obvious interference comes from the Raman scattering signal arising from the surrounding medium and/or from contaminants that are also adsorbed at the enhancing surface, as was illustrated for metal colloids. In addition, there are important variables to be carefully monitored in SERS experiments conducted with a micro-Raman type of system. In micro-Raman spectroscopy, the surface sampled usually is between 500 and $1\,\mu m^2$. The powers at the sample are on the order of 10 mW–1 nW, which gives

Table 6.1 Laser power and energy density for 633 nm laser line and a 50× microscope objective

%	mW	$W\ cm^{-2}$	Photons $s^{-1}\ \mu m^{-2}$
100	4.5	6.49×10^5	1.43×10^{16}
50	2.250	2.24×10^5	7.17×10^{15}
10	0.45	6.49×10^4	1.43×10^{15}
1	0.045	6.49×10^3	1.43×10^{14}
0.5	0.0225	2.24×10^3	7.17×10^{13}
0.025	0.001125	1.12×10^2	3.58×10^{12}

energy densities (ED) of 10^6–10^3 W cm^{-2}. A somewhat better gage in micro-Raman spectroscopy is the quantity of photon flux [3]. Both ED and photon flux are illustrated in Table 6.1. Although the ED is a concern, micro-Raman spectroscopy's benefits of high spatial resolution and control over sample volume and probed surface more than offset it. In addition, owing to the enhanced intensity in SERS experiments and the increased sensitivity of detectors, large signal-to-noise ratios are easily achieved with small laser powers. An important element of the collection optics is the numerical aperture (*NA*) of the objective being used. This attribute is a measure of the collection solid angle of an objective and, in general, the higher is *NA*, the better is the collection efficiency. There may also be a connection between the *NA* of the objective used and the measured empirical enhancement factor. A study by Wei *et al.* of SERS using ordered gold particle arrays [126] has shown a dependence between the *NA* of the objective used and the relative enhancement factor, with larger *NA*s providing larger enhancements. The explanation put forth is that the angle of observation is different for different *NA* values. Low values have a more obtuse half-angle (*NA* 0.25–15°) whereas larger values are more oblique (*NA* 0.75–49°). Raman cross-sections in general have a polarization dependence and so do the enhancing metal substrates. Two recent studies clearly demonstrate that the polarization of exciting fields leads to polarization of the enhancement. Jeong *et al.* [116] measured the polarization dependence of SERS for highly ordered nanowire rafts. They established a correlation between enhancements for those wires aligned parallel and perpendicular to the polarization of excitation. In a study using aggregated colloids by Xu and Kaell [127], a similar polarization dependence was demonstrated. Calculations showed that the region of interstitial sites is strongly polarized along the axis of interactions and hence excitation aligned with this axis offers maximum enhancement. They also indicated that for clusters made up of more than two particles, this dependence is more isotropic. It would seem, then, that for highly ordered substrates of interacting nanoparticles, polarization of the excitation line becomes increasingly important and for optimal results it is another factor that needs to be accounted for. A further issue that is a concern in any Raman experiment is that of local heating induced by the laser. However, with the low powers utilized in SERS experiments, the heating is minimal. Measured [128] and calculated [19] laser heating data show very small increases in temperatures for the energy densities typically utilized in SERS experiments. Weiss and Haran [19] calculated that for an illumination of 100 W cm^{-2}, there is a less than 0.1 K increase in a nanoparticle's temperature.

For those who use continuous extended scanning rather then the traditional static scans, it should be noted that large ranges (3500–100 cm^{-1} with accumulation times of 10 s or greater), the sample can undergo a photo-induced change during the time of the scan. This can result in a spectrum where the high-and low-wavenumber regions correspond to two different chemical systems.

6.8 APPLICATIONS AND OUTLOOK

SERS/SERRS is an analytical technique that is expanding its realm of applications from chemical–biochemical analysis to nanostructure characterization and biological and biomedical applications. The potential of surface-enhanced spectroscopy for nanostructure characterization is a new development and is a field in itself [129]. With an understanding of the basic experimental requirements to achieve SERS/SERRS, and the clear theoretical guidance provided by the electromagnetic interpretation, the application of SERS and SERRS continues to grow at a fast pace. An important caveat for practitioners is to take advantage of the important distinction between average SERS and ultrasensitive SERS, to avoid inaccuracy in the interpretation of the results. The edge is in the synergy of the wealth of molecular specific data provided by vibrational spectroscopy with extremely high spatial resolution.

Since there is a enormous amount of published work, only a few examples will be given here to illustrate the wide scope of applications. Metal colloids are by far the most common SERS-active substrate used in analytical applications. Dou and Ozaki [130] reviewed the use of metal colloids for analytical SERS of biological molecules. Kneipp *et al.* [131] discussed in their review the opportunities wrought by SERS in the field of biophysics and biomedical spectroscopy, relevant to the detection of molecules and processes. Vo-Dinh [132] provided a review of SERS in practical applications in environmental and biomedical areas, pointing to new developments in fiber-optic SERS monitors, SERS nanoprobes near-field SERS probes and SERS-based bioassays. Pinzaru *et al.* [9] described applications of Raman and SERS towards fundamental structural investigation and 'limit of detection' analysis of several widely used pharmaceutical compounds, including discussions of quantitative analysis, drug–excipient interaction and pH dependence. Grow *et al.* [133] reviewed the new biochip technology for label-free detection of pathogens and their toxins using SERS. Possible SERS applications in forensic science, including aspirin tablets, drugs, explosives, lipsticks, shoe polishes, fibers and printing inks, have been reviewed by Smith *et al.* [10].

The trend to develop SERS as a rapid whole-organism fingerprinting method for the characterization of bacteria has been illustrated by Jarvis and Goodacre [134]. They studied the detection of the bacteria associated with urinary tract infection (UTI), which affects millions of people across all age groups. The organisms most commonly found to be the cause of UTI are members of the family Enterobacteriaceae, particularly *Escherichia coli*, which is the causal agent in more than 50% of cases. Farquharson *et al.* [135] reported the SERS detection of dipicolinic acid extracted from *Bacillus* spores, followed by the analysis of the chemotherapy drug 5-fluorouracil in saliva. They used silver-doped sol–gel as a SERS-active substrate, reporting that 5-fluorouracil and physiological thiocyanate produced SERS, whereas large biochemicals, such as enzymes and proteins, did not [136].

SERS/SERRS has been shown to be a powerful analytical identification method when used in conjunction with any of the separation science techniques. For instance, Sagmuller *et al.* [137] developed an analytical protocol based on high-performance liquid chromatography (HPLC) in combination with SERS as a detection technique for the reliable identification of the ingredients of illicitly sold drugs or other pharmaceutical compounds such as cocaine, heroin and amphetamines, and the pharmaceuticals (nor-)papaverine and procaine. The technique was later adapted for the analysis of drugs in human blood and urine [138]. The practicality of an at-line capillary electrophoresis (CE)–SERRS configuration has been reported [139]. The CE effluent was deposited on a moving thin-layer chromatographic (TLC) plate. Three acidic dyes (Food Yellow 3, Acid Orange 7 and Food Red 1) were tested. Silver sol was used as the SERS substrate and the sample was excited with 514.5 nm radiation from an argon ion laser. Analyte identification could be achieved down to a few picomoles deposited on the TLC plate. Nirode *et al.* [140] have demonstrated the feasibility of the on-column SERS detection in CE using running buffers that contain silver colloidal solutions. The effects of laser power, wavelength, spectral acquisition time, silver colloidal concentration and applied voltage (i.e. flow rate) on the quality of SERS spectra were discussed.

Ayora Canada *et al.* [141] developed a fully automated flow system comprising a dedicated flow cell, which contains two inlet and two outlet channels. The beads carrying cation-exchange moieties are retained in the flow cell and subsequently treated with silver nitrate and hydroxylamine solution, forming a SERS-active silver layer on the beads. The analyte is then introduced from the second inlet channel such that the interaction between the activated SERS beads and analyte occur within the focus of the laser excitation beam. Quantitative studies were carried out for 9-aminoacridine and acridine, showing linear responses of 1–100 and 50–1000 nmol L^{-1}, respectively.

These few examples give the flavor of the realm of applications of SERS/SERRS as an analytical tool. The inroads of SERS in bioscience and nano-bioscience are particularly impressive and are clearly the fastest growing branch of the field. The recent measurements of enzyme activities using SERRS exemplifies the potential of indirect measurements and applications to *in vivo* detection with a high degree of selectivity [142]. The aim of this chapter was to illustrate the experimental variables that need to be considered when selecting and fabricating the SERS-active substrate and the control of variables available to the experimenter during the acquisition of the SERS spectra. After a SERS/SERRS spectrum has been recorded, special attention should be paid to spectral interpretation of the SERS/SERRS results as detailed in Chapter 4. Future work, to aid the systematic application of SERS/SERRS, entails a logical description of these factors for a given substrate. Finally, the compilation of a SERS/SERRS spectral library that takes such experimental conditions into account will be a tremendous asset in the daily use of these powerful analytical methods in both industrial and research laboratories.

REFERENCES

[1] D.S. Kliger, J.W. Lewis and C.E. Randall, *Polarized Light in Optics and Spectroscopy*, Academic Press, Boston, 1990.
[2] C.G. Enke, The Art and Science of Chemical Analysis, John Wiley & Sons, Inc., New York, 2001.

[3] R.L. McCreery, *Raman Spectroscopy for Chemical Analysis*, John Wiley & Sons, Inc., New York, 2000.

[4] G. Fini, Applications of Raman spectroscopy to pharmacy, *J. Raman Spectrosc.* 2004, **35**, 335–337.

[5] E. Koglin and J.M. Sequaris, Surface enhanced Raman scattering of biomolecules, *Top. Curr. Chem.* 1986, **134**, 1–57.

[6] L.A. Lyon, C.D. Keating, A.P. Fox, B.E. Baker, L. He, S.R. Nicewarner, S.P. Mulvaney and M.J. Natan, Raman spectroscopy, *Anal. Chem.* 1998, **70**, 341R–361R.

[7] I. Nabiev, I. Chourpa and M. Manfait, Applications of Raman and surface-enhanced Raman scattering spectroscopy in medicine, *J. Raman Spectrosc.* 1994, **25**, 13–23.

[8] R.F. Paisley and M.D. Morris, Surface enhanced Raman spectroscopy of small biological molecules, *Prog. Anal. Spectrosc.* 1988, **11**, 111–140.

[9] S.C. Pinzaru, I. Pavel, N. Leopold and W. Kiefer, Identification and characterization of pharmaceuticals using Raman and surface-enhanced Raman scattering, *J. Raman Spectrosc.* 2004, **35**, 338–346.

[10] W.E. Smith, P.C. White, C. Rodger and G. Dent, Raman and surface enhanced resonance Raman scattering: applications in forensic science, *Pract. Spectrosc.* 2001, **28**, 733–748.

[11] T. Vo-Dinh, Remote monitors for *in situ* characterization of hazardous wastes, *Waste Manage. Ser.* 2004, **4**, 485–502.

[12] K. Kneipp, H. Kneipp, I. Itzkan, R.R. Dasari and M.S. Feld, Ultrasensitive chemical analysis by Raman spectroscopy, *Chem. Rev.* 1999, **99**, 2957.

[13] P.J.G. Goulet, N.P.W. Pieczonka and R.F. Aroca, Overtones and combinations in single-molecule surface-enhanced resonance Raman scattering spectra, *Anal. Chem.* 2003, **75**, 1918–1923.

[14] S. Nie and S.R. Emory, Probing single molecules and single nanoparticles by surface-enhanced Raman scattering, *Science* 1997, **275**, 1102–1106.

[15] H.I.S. Nogueira, J.J.C. Teixeira-Dias and T. Trindade, Nanostructured metals in surface enhanced Raman spectroscopy, *Encycl. Nanosci. Nanotechnol.* 2004, **7**, 699–715.

[16] W.E. Doering and S. Nie, Single-molecule and single-nanoparticle SERS: examining the roles of surface active sites and chemical enhancement, *J. Phys. Chem. B* 2002, **106**, 311–317.

[17] W.E. Doering and S. Nie, Spectroscopic tags using dye-embedded nanoparticles and surface-enhanced Raman scattering, *Anal. Chem.* 2003, **75**, 6171–6176.

[18] A. Otto, A. Bruckbauer and Y.X. Chen, On the chloride activation in SERS and single molecule SERS, *J. Mol. Struct.* 2003, **661–662**, 501–514.

[19] A. Weiss and G.J. Haran, Time-dependent single-molecule Raman scattering as a probe of surface dynamics, *J. Phys. Chem. B* 2001, **105**, 12348–12345.

[20] J.C. Tully, Chemical dynamics at metal surfaces, *Annu. Rev. Phys. Chem.* 2000, **51**, 153–178.

[21] S. Schneider, P. Halbig, H. Grau, and U. Nickel, Reproducible preparation of silver sols with uniform particle size for application in surface-enhanced Raman spectroscopy, *Photochem. and Photobiology*, 1994, **60**, 605–610.

[22] D.L. Feldheim and C.A. Foss (eds), *Metal Nanoparticles. Synthesis, Characterization and Applications*, Marcel Dekker, New York, 2002.

[23] M. Faraday, *Philos. Trans. R. Soc. London*, 1857, **147**, 145.

[24] J.A. Creighton, C.G. Blatchford and M.G. Albrecht, Plasma resonance enhancement of Raman scattering by pyridine adsorbed on silver or gold sol particles of size comparable to the excitation wavelength, *J. Chem. Soc., Faraday Trans. 2*, 1979, **75**, 790–798.

[25] M. Kerker, O. Siiman, L.A. Bumm and D.S. Wang, Surface enhanced Raman scattering (SERS) of citrate ion adsorbed on colloidal silver, *Appl. Opt.* 1980, **19**, 3253–3255.

[26] P.C. Lee and D. Meisel, Adsorption and surface-enhanced Raman of dyes on silver and gold sols, *J. Phys. Chem.* 1982, **86**, 3391–3395.

[27] S. Sanchez-Cortes, J.V. Garcia-Ramos and G. Morcillo, Morphological study of metal colloids employed as substrate in SERS spectroscopy, *J. Colloid Interface Sci.* 1994, **167**, 428–436.

[28] S. Sanchez-Cortes, J.V. Garcia-Ramos, G. Morcillo and A. Tinti, Morphological study of silver colloids employed in SERS: activation when exciting in visible and infrared regions, in *Spectroscopy of Biological Molecules, 6th European Conference on the Spectroscopy of Biological Molecules, Villeneuve d'Ascq, France, September 3–8, 1995*, pp. 29–30.

[29] S. Sanchez-Cortes and J.V. Garcia-Ramos, SERS of AMP on different silver colloids, *J. Mol. Structu.* 1992, **274**, 33–45.

[30] O. Siiman, A. Lepp and M. Kerker, Combined surface-enhanced and resonance-Raman scattering from the aspartic acid derivative of methyl orange on colloidal silver, *J. Phys. Chem.* 1983, **87**, 5319–5325.

[31] P. Matejka, The role of surfactants in the surface enhanced Raman scattering spectroscopy, Chem. Listy 1982, **86**, 875–883.

[32] P. Matejka, B. Vlckova, J. Vohlidal, P. Pancoska and V. Baumruk, The role of Triton X-100 as an adsorbate and a molecular spacer on the surface of silver colloid: a surface-enhanced Raman scattering study, *J. Phys. Chem.* 1992, **96**, 1361–1366.

[33] Y. Shao, Y. Jin and S. Dong, Synthesis of gold nanoplates by aspartate reduction of gold chloride, *Chem. Commun.* 2004, 1104–1105.

[34] S. Chen and D.L. Carroll, Synthesis and characterization of truncated triangular silver nanoplates, *Nano Lett.* 2002, **2**, 1003–1007.

[35] Y. Sun and Y. Xia, Shape-controlled synthesis of gold and silver nanoparticles, *Science* 2002, **298**, 2176–2179.

[36] J.B. Liu, W. Dong, P. Zhan, S.Z. Wang, J.H. Zhang and Z.L. Wang, Synthesis of bimetallic nanoshells by an improved electroless plating method, *Langmuir* 2005, **21**, 1683–1686.

[37] Y.-J. Zhu and X.-L. Hu, Microwave-polyol preparation of single-crystalline gold nanorods and nanowires, *Chem. Lett.* 2003, **32**, 1140–1141.

[38] D.S. Dos Santos, Jr, P.J.G. Goulet, N.P.W. Pieczonka, O.N. Oliveira Jr and R.F. Aroca, Gold nanoparticle embedded, self-sustained chitosan films as substrates for surface-enhanced Raman scattering, *Langmuir* 2004, **20**, 10273–10277.

[39] R.F. Aroca, P.J.G. Goulet, D.S. dos Santos Jr, R.A. Alvarez-Puebla and O.N. Oliveira Jr, Silver nanowire layer-by-layer films as substrates for surface-enhanced Raman scattering, *Anal. Chem.* 2005, **77**, 378–382.

[40] S. Sanchez-Cortes and J.V. Garcia-Ramos, Anomalous Raman bands appearing in surface-enhanced Raman spectra, *J. Raman Spectrosc.* 1998, **29**, 365–371.

[41] A. Otto, What is observed in single molecule SERS, and why?, *J. Raman Spectrosc.* 2002, **33**, 593–598.

[42] J.C. Tsang, J.E. Demuth, P.N. Sanda and J.R. Kirtley, Enhanced Raman scattering from carbon layers on silver, *Chem. Phys. Lett.* 1980, **76**, 54–57.

[43] A.C. Ferrari and J. Robertson, Resonant Raman spectroscopy of disordered, amophous, and diamondlike carbon, *Phys. Rev. B* 2001, **64**, 1–13.

[44] C.E. Taylor, S.D. Garvey and J.E. Pemberton, Carbon contamination at silver surfaces: surface preparation procedures evaluated by Raman spectroscopy and X-ray photoelectron spectroscopy, *Anal. Chem.* 1996, **68**, 2401–2408.

[45] B. Pettinger and A. Kudelski, SERS on carbon chain segments: monitoring locally surface chemistry, *Chem. Phys. Lett.* 2000, **321**, 356–362.

[46] E.J. Bjerneld, F. Svedberg, P. Johansson and M. Kaell, Direct observation of heterogeneous photochemistry on aggregated Ag nanocrystals using Raman spectroscopy: the case of photoinduced degradation of aromatic amino acids, *J. Phys. Chem. A* 2004, **108**, 4187–4193.

[47] P. Etchegoin, H. Liem, R.C. Maher, L.F. Cohen, R.G.C. Brown, H. Hartigan, M.J.T. Milton and G.C. Gallop, A novel amplification mechanism for surface enhanced Raman scattering, *Chem. Phys. Lett.* 2002, **366**, 115–121.

[48] A. Kudelski and B. Pettinger, Fluctuations of surface-enhanced Raman spectra of CO adsorbed on gold substrates, *Chem. Phys. Lett.* 2004, **383**, 76–79.

[49] J. Neddersen, G. Chumanov and T.M. Cotton, Laser ablation of metals: a new method for preparing SERS active colloids, *Appl. Spectrosc.* 1993, **47**, 1959–1964.

[50] M. Prochazka, P. Mojzes, J. Stepanek, B. Vlckova and P.-Y. Turpin, Probing applications of laser-ablated Ag colloids in SERS spectroscopy: improvement of ablation procedure and SERS spectral testing, *Anal. Chem.* 1997, **69**, 5103–5108.

[51] D. Myers, *Surfaces, Interfaces, and Colloids*, John Wiley & Sons, Inc., New York, 1999.

[52] P.C. Hiemenz and R. Rajagopalan, *Principles of Colloid and Surface Chemistry*, Marcel Dekker, New York, 1997.

[53] S. Lecomte, P. Matejka and M.H. Baron, Correlation between surface enhanced Raman scattering and absorbance changes in silver colloids. Evidence for the chemical enhancement mechanism, *Langmuir* 1998, **14**, 4373–4377.

[54] K. Faulds, R.E. Littleford, D. Graham, G. Dent and W.E. Smith, Comparison of surface-enhanced resonance Raman scattering from unaggregated and aggregated nanoparticles, *Anal. Chem.* 2004, **76**, 592–598.

[55] G. Levi, J. Pantigny, J.P. Marsault and J. Aubard SER spectra of acridine and acridinium ions in colloidal silver sols. Electrolytes and pH effects, *J. Raman Spectrosc.* 1993, **24**, 745–752.

[56] J.C. Maxwell-Garnet, *Trans. R. Soc. London* 1904, **203**, 384.

[57] G. Mie, Considerations on the optics of turbid media, especially colloidal metal sols, *Ann. Phys.* 1908, **25**, 377–445.

[58] U. Kreibig, *Optical Properties of Metal Clusters*, Springer-Verlag, New York, 1995.

[59] C.F. Bohren and D.R. Huffman, *Absorption and Scattering of Light by Small Particles*, John Wiley & Sons, Inc., New York, 1983.

[60] M. Kerker, *The Scattering of Light and Other Electromagnetic Radiation*, Academic Press, New York, 1969.

[61] S. Yoshida, T. Yamaguchi and A. Kimbara, Optical properties of aggregated silver films, *J. Opt. Soc. Am.* 1971, **61**, 62–69.

[62] C.G. Granqvist and O. Hunderi, Optical properties of ultrafine gold particles, *Phys. Rev. B* 1977, **16**, 3513.

[63] G.C. Papavassiliou, *Prog. Solid State Chem.* 1984, **12**, 185–271.
[64] D.J. Semin and K.L. Rowlen, Influence of vapor deposition parameters on SERS active Ag film morphology and optical properties, *Anal. Chem.* 1994, **66**, 4324–4331.
[65] K.L. Norrod, L.M. Sudnik, D. Rousell and K.L. Rowlen, Quantitative comparison of five SERS substrates: sensitivity and limit of detection, *Appl. Spectrosc.* 1997, **51**, 994–1001.
[66] D.L. Smith, *Thin-film Deposition. Principle and Practice*, McGraw-Hill, New York, 1995.
[67] R. Aroca, C. Jennings, G.J. Kovac, R.O. Loutfy and P.S. Vincett, Surface-enhanced Raman scattering of Langmuir–Blodgett monolayers of phthalocyanine by indium and silver island films, *J. Phys. Chem.* 1985, **89**, 4051–4054.
[68] R. Aroca and G.J. Kovacs, *Surface enhanced Raman scattering on silver-coated tin spheres, J. Mol. Structure*, 1988, **174**, 53–58.
[69] E. Vogel and W. Kiefer, Investigation of SERS-activity of binary Ag–Ni films, *Asian J. Phys.* 2000, **9**, 841–849.
[70] N.T. Flynn and A.A. Gewirth, Attenuation of surface-enhanced spectroscopy response in gold–platinum core-shell nanoparticles, *J. Raman Spectrosc.* 2002, **33**, 243–251.
[71] M. Kerker (ed.), *Papers on Surface-enhanced Raman Scattering*, SPIE, Bellingham, WA, 1990.
[72] S.E. Roark and K.L. Rowlen, Thin Ag films: influence of substrate and postdeposition treatment on morphology and optical properties, *Anal. Chem.* 1994, **66**, 261–270.
[73] J.L. Martinez, Y. Gao, T. Lopez-Rios and A. Wirgin, Anisotropic surface-enhanced Raman scattering at obliquely evaporated silver films, *Phys. Rev. B* 1987, **35**, 9481–9488.
[74] E.A. Wachter, A.K. Moore and J.W. Haas III, Fabrication of tailored needle substrates for surface-enhanced Raman scattering, *Vib. Spectrosc.* 1992, **3**, 73–78.
[75] T.R. Jensen, M.D. Malinsky, C.L. Haynes and R.P. Van Duyne, Nanosphere lithography: tunable localized surface plasmon resonance spectra of silver nanoparticles, *J. Phys. Chem. B* 2000, **104**, 10549–10556.
[76] A. Otto, The electronic contribution to SERS. The present experimental status, *Colloids Surf.* 1989, **38**, 27–36.
[77] A. Otto, I. Mrozek, H. Grabhorn and W. Akemann, Surface-enhanced Raman scattering, *J. Phys.: Condens. Matter* 1992, **4**, 1143–1212.
[78] E.V. Albano, S. Daiser, R. Miranda and K. Wandelt, On the porosity of coldly condensed SERS active films. II. Comparison of adsorption and Raman scattering of pyridine, *Surf. Sci.* 1985, **150**, 386–398.
[79] R.K. Chang and B.L. Laube, Surface-enhanced Raman scattering and nonlinear optics applied to electrochemistry, *Crit. Rev. Solid State Mater. Sci.* 1984, **12**, 1–73.
[80] D.V. Murphy, K.U. Von Raben, R.K. Chang and P.B. Dorain, Surface-enhanced hyper-Raman scattering from sulfite ion adsorbed on silver powder, *Chem. Phys. Lett.* 1982, **85**, 43–47.
[81] H. Luo and M.J. Weaver, surface-enhanced Raman scattering as a versatile vibrational probe of transition-metal interfaces: thiocyanate coordination modes on platinum-group versus coinage-metal electrodes, *Langmuir* 1999, **15**, 8743–8749.

[82] M.J. Weaver, Raman and infrared spectroscopies as *in situ* probes of catalytic adsorbate chemistry at electrochemical and related metal–gas interfaces: some perspectives and prospects, *Top. Catal.* 1999, **8**, 65–73.

[83] Z.-Q. Tian, B. Ren and D.-Y. Wu, Surface-enhanced Raman scattering: from noble to transition metals and from rough surfaces to ordered nanostructures, *J. Phys. Chem. B* 2002, **106**, 9463–9483.

[84] B. Ren, J.L. Yao, C.X. She, Q.J. Huang and Z.Q. Tian, Surface Raman spectroscopy on transition metal surfaces, *Internet J. Vib. Spectrosc.* [online computer file] 2000, **4**, (no pages given).

[85] Z.-Q. Tian and B. Ren, Raman spectroscopy of electrode surfaces, *Encyclo. Electrochem.* 2003, **3**, 572–659.

[86] S.A. Wasileski, S. Zou and M.J. Weaver, Surface-enhanced Raman scattering from substrates with conducting or insulator overlayers: electromagnetic model predictions and comparisons with experiment, *Appl. Spectrosc.* 2000, **54**, 761–772.

[87] Z.-Q. Tian and B. Ren, Adsorption and reaction at electrochemical interfaces as probed by surface-enhanced Raman spectroscopy, *Ann. Rev. Phys. Chem.* 2004, **55**, 197–229.

[88] D.H. Murgida, P. Hildebrandt, J. Wei, Y.F. He, H. Liu and D.H. Waldeck, Surface-enhanced resonance raman spectroscopic and electrochemical study of cytochrome c bound on electrodes through coordination with pyridinyl-terminated self-assembled monolayers, *J. Phys. Chem. B* 2004, **108**, 2261–2269.

[89] V. Oklejas and J.M. Harris, Potential-dependent surface-enhanced Raman scattering from adsorbed thiocyanate for characterizing silver surfaces with improved reproducibility, *Appl. Spectrosc.* 2004, **58**, 945–951.

[90] A.G. Brolo, P. Germain and G. Hager, Investigation of the adsorption of L-cysteine on a polycrystalline silver electrode by surface-enhanced Raman scattering (SERS) and surface-enhanced second harmonic generation (SESHG), *J. Phys. Chem. B* 2002, **106**, 5982–5987.

[91] P. Corio, P., M.L.A. Temperini, P.S. Santos and J.C. Rubim, Contribution of the charge transfer mechanism to the surface-enhanced Raman scattering of the binuclear ion complex $[Fe_2(Bpe)(CN)_{10}]^{6-}$ adsorbed on a silver electrode in different solvents, *Langmuir* 1999, **15**, 2500–2507.

[92] A.C. Sant'Ana, P.S. Santos and M.L.A. Temperini, The adsorption of squaric acid and its derived species on silver and gold surfaces studied by SERS, *J. Electroanal. Chem.* 2004, **571**, 247–254.

[93] J. Rodrigues de Sousa, M.M.V. Parente, I.C.N. Diogenes, L.G.F. Lopes, P. de Lima Neto, M.L.A. Temperini, A.A. Batista and I.D.S. Moreira, A correlation study between the conformation of the 1,4-dithiane SAM on gold and its performance to assess the heterogeneous electron-transfer reactions. *J. Electroanal. Chem.* 2004, **566**, 443–449.

[94] B. Ren, X.-F. Lin, Z.-L. Yang, G.-K. Liu, R.F. Aroca, B.-W. Mao and Z.-Q. Tian, Surface-enhanced Raman scattering in the ultraviolet spectral region: UV–SERS on rhodium and ruthenium electrodes, *J. Am. Chem. Soc.* 2003, **125**, 9598–9599.

[95] K. Kneipp, Y. Wang, H. Kneipp, R.R. Dasari and M.S. Feld, An approach to single molecule detection using surface-enhanced Raman scattering (SERS), *Exp. Tech. Phys.* 1995, **41**, 225–234.

[96] K. Kneipp, Y. Wang, H. Kneipp, L.T. Perelman, I. Itzkan, R.R. Dasari and M.S. Feld, Single molecule detection using surface-enhanced Raman scattering (SERS), *Phys. Rev. Lett.* 1997, **78**, 1667–1670.

[97] M. Futamata, Single molecule sensitivity in SERS: importance of junction of adjecent Ag nanoparticles. *Faraday Discuss.* 2005, **132**, 1–17.

[98] T. Basche, W.E. Moerner, M. Orrit and U.P. Wild (eds), *Single-molecule Detection, Optical Detection, Imaging and Spectroscopy*, VCH, Weinhem, 1997.

[99] P. Tamarat, A. Maali, B. Lounis and M. Orrit, Ten years of single-molecule spectroscopy, *J. Phys. Chem. A* 2000, **104**, 1–16.

[100] W.E. Moerner, A dozen years of single-molecule spectroscopy in physics, chemistry, and biophysics, *J. Phys. Chem. B* 2002, **106**, 910–927.

[101] W.E. Moerner, Thirteen years of single-molecule spectroscopy in physical chemistry and biophysics, *Springer Ser. Chem. Phys.* 2001, **67**, 32–61.

[102] A.M. Michaels, J. Jiang and L. Brus, Ag nanocrystal junctions as the site for surface-enhanced Raman scattering of single Rhodamine 6G molecules, *J. Phys. Chem. B* 2000, **104**, 11965–11971.

[103] A.M. Michaels, M. Nirmal and L.E. Brus, Surface enhanced Raman spectroscopy of individual Rhodamine 6G molecules on large Ag nanocrystals, *J. Am. Chem. Soc.* 1999, **121**, 9932–9939.

[104] C. Eggeling, J. Schaffer, C.A. Seidel, J. Korte, G. Brehm, S. Schneider and W. Schrof, Homogeneity, transport, and signal properties of single Ag particles studied by single-molecule surface-enhanced resonance Raman scattering, *J. Phys. Chem. A* 2001, **105**, 3673–3679.

[105] A. Weiss and G. Haran, Time-dependent single-molecule Raman scattering as a probe of surface dynamics, *J. Phys. Chem. B* 2001, **105**, 12348–12354.

[106] S.R. Emory, W.P. Ambrose, P.M. Goodwin and R.A. Keller, Observing single-molecule chemical reactions on metal nanoparticles, *Proc. SPIE* 2001, **4258**, 63–72.

[107] H. Xu, E.J. Bjerneld, M. Kall and L. Borjesson, Spectroscopy of single hemoglobin molecules by surface enhanced Raman scattering, *Phys. Rev. Lett.* 1999, **83**, 4357–4360.

[108] K. Kneipp, H. Kneipp, I. Itzkan, R.R. Dasari and M.S. Feld, Single molecule detection using near infrared surface-enhanced Raman scattering, *Springer Ser. Chem. Phys.* 2001, **67**, 144–160.

[109] A.R. Bizzarri and S. Cannistraro, Surface-enhanced resonance Raman spectroscopy signals from single myoglobin molecules, *Appl. Spectrosc.* 2002, **56**, 1531–1537.

[110] E.J. Bjerneld, Z. Foeldes-Papp, M. Kaell and R. Rigler, Single-molecule surface-enhanced Raman and fluorescence correlation spectroscopy of horseradish peroxidase, *J. Phys. Chem. B* 2002, **106**, 1213–1218.

[111] P. Etchegoin, H. Liem, R.C. Maher, L.F. Cohen, R.J.C. Brown, H. Hartigan, M.J.T. Milton and J.C. Gallop, A novel amplification mechanism for surface enhanced Raman scattering, *Chem. Phys. Lett.*, 2002, **366**, 115–121.

[112] A.R. Bizzarri and S. Cannistraro, Temporal fluctuations in the SERRS spectra of single iron-protoporphyrin IX molecule, *Chem. Phys.* 2003, **290**, 297–306.

[113] S. Habuchi, M. Cotlet, R. Gronheid, G. Dirix, J. Michiels, J. Vanderleyden, F.C. De Schryver and J. Hofkens, Single-molecule surface enhanced resonance Raman spectroscopy of the enhanced green fluorescent protein, *J. Am. Chem. Soc.* 2003, **125**, 8446–8447.

[114] T. Vosgroene and A.J. Meixner, Surface- and resonance-enhanced micro-Raman spectroscopy of xanthene dyes: from the ensemble to single molecules, *ChemPhysChem* 2005, **6**, 154–163.
[115] P. Etchegoin, L.F. Cohen, H. Hartigan, R.J.C. Brown, M.J.T. Milton and J.C. Gallop, Localized plasmon resonances in inhomogeneous metallic nanoclusters, *Chem. Phys. Lett.* 2004, **383**, 577–583.
[116] D.H. Jeong, Y.X. Zhang and M. Moskovits, Polarized surface enhanced Raman scattering from aligned silver nanowire rafts, *J. Phys. Chem. B* 2004, **108**, 12724–12728.
[117] M. Mandal, N.R. Jana, S. Kundu, S.K. Ghosh, M. Panigrahi and T. Pal, Synthesis of Au core–Ag shell type bimetallic nanoparticles for single molecule detection in solution by SERS method, *J. Nanoparticle Res.* 2004, **6**, 53–61.
[118] B. Pettinger, G. Picardi, R. Schuster and G. Ertl, Surface-enhanced and STM–tip-enhanced Raman spectroscopy at metal surfaces, *Single Mol.* 2002, **3**, 285–294.
[119] C.J.L. Constantino, T. Lemma, P.A. Antunes and R. Aroca, Single-molecule detection using surface-enhanced resonance Raman scattering and Langmuir–Blodgett monolayers, *Anal. Chem.* 2001, **73**, 3674–3678.
[120] P.J.G. Goulet, N.P.W. Pieczonka and R.F. Aroca, Overtones and combinations in single-molecule surfaced-enhance resonance Raman scattering spectra, *Anal. Chem.* 2003, **75**, 1918–1923.
[121] R.C. Maher, M. Dalley, E.C. Le Ru, L.F. Cohen, P.G. Etchegoin, H. Hartigan, R.J.C. Brown and M.J.T. Milton, Physics of single molecule fluctuations in surface enhanced Raman spectroscopy active liquids, *J. Chem. Phys.* 2004, **121**, 8901–8910.
[122] M. Lippitz, F. Kulzer and M. Orrit, Statistical evaluation of single nano-object fluorescence, *ChemPhysChem* 2005, **6**, 770–789.
[123] M.L. Jacobson and K.L. Rowlen, Photodynamics on thin silver films, *Chem. Phys. Lett.* 2005, **401**, 52–57.
[124] H.P. Lu, Site-specific Raman spectroscopy and chemical dynamics of nanoscale interstitial systems, *J. Phys. Condens. Matter* 2005, **17**, R333–R355.
[125] A.J. Meixner, T. Vosgrone and M. Sackrow, Nanoscale surface-enhanced Raman scattering spectroscopy of single molecules on isolated silver clusters, *J. Lumin.* 2001, **94–95**, 147–152.
[126] A. Wei, B. Kim, B. Sadtler and S.L. Tripp, Tunable surface-enhanced Raman scattering from large gold nanoparticle arrays, *ChemPhysChem* 2001, **2**, 743–745.
[127] H. Xu and M. Kaell, Polarization-dependent surface-enhanced Raman spectroscopy of isolated silver nanoaggregates, *ChemPhysChem* 2003, **4**, 1001–1005.
[128] S. Xie, M.P. Rosynek and J.H. Lunsford, Effects of laser heating on the local temperature and composition in Raman spectroscopy: a study of $Ba(NO_3)_2$ and BaO_2 decompostion. *Appl. Spectrosc.* 1999, **53**, 1183–1187.
[129] D. Roy and J. Fendler, Reflection and absorption techniques for optical characterization of chemically assembled nanomaterials, *Adv. Materi.* 2004, **16**, 479–508.
[130] X.-M. Dou and Y. Ozaki, Surface-enhanced Raman scattering of biological molecules on metal colloids: basic studies and applications to quantitative assay, *Rev. Anal. Chem.* 1999, **18**, 285–321.
[131] K. Kneipp, H. Kneipp, I. Itzkan, R.R. Dasari and M.S. Feld, Surface-enhanced Raman scattering and biophysics, *J. Phys.: Condens. Matter* 2002, **14**, R597–R624.
[132] T. Vo-Dinh and D.L. Stokes, Surface-enhanced Raman scattering (SERS) for biomedical diagnostics, *Biomed. Photonics Handb.* 2003, 64/1–64/39.

[133] A.E. Grow, L.L. Wood, J.L. Claycomb and P.A. Thompson, New biochip technology for label-free detection of pathogens and their toxins, *J. Microbiol. Methods* 2003, **53**, 221–233.
[134] R.M. Jarvis and R. Goodacre, Discrimination of bacteria using surface-enhanced Raman spectroscopy, *Anal. Chem.* 2004, **76**, 40–47.
[135] S. Farquharson, A.D. Gift, P. Maksymiuk and F.E. Inscore, Rapid dipicolinic acid extraction from *Bacillus* spores detected by surface-enhanced Raman spectroscopy, *Appl. Spectrosc.* 2004, **58**, 351–354.
[136] S. Farquharson, C. Shende, F.E. Inscore, P. Maksymiuk and A.D. Gift, Analysis of 5-fluorouracil in saliva using surface-enhanced Raman spectroscopy, *J. Raman Spectrosc.* 2005, **36**, 208–212.
[137] B. Sagmuller, B. Schwarze, G. Brehm, G. Trachta and S. Schneider, Identification of illicit drugs by a combination of liquid chromatography and surface-enhanced Raman scattering spectroscopy, *J. Mol. Struct.* 2003, **661–662**, 279–290.
[138] G. Trachta, B. Schwarze, B. Saegmuller, G. Brehm and S. Schneider, Combination of high-performance liquid chromatography and SERS detection applied to the analysis of drugs in human blood and urine, *J. Mol. Struct.* 2004, **693**, 175–185.
[139] R.M. Seifar, R.J. Dijkstra, A. Gerssen, F. Ariese, U.A.T. Brinkman and C. Gooijer, At-line coupling of capillary electrophoresis and surface-enhanced resonance Raman spectroscopy, *J. Sep. Sci.* 2002, **25**, 814–818.
[140] W.F. Nirode, G.L. Devault, M.J. Sepaniak and R.O. Cole, On-column surface-enhanced Raman spectroscopy detection in capillary electrophoresis using running buffers containing silver colloidal solutions, *Anal. Chem.* 2000, **72**, 1866–1871.
[141] M.J. Ayora Canada, A. Ruiz Medina, J. Frank and B. Lendl, Bead injection for surface enhanced Raman spectroscopy: automated on-line monitoring of substrate generation and application in quantitative analysis, *Analyst* 2002, **127**, 1365–1369.
[142] B.D. Moore, L. Stevenson, A. Watt, S. Flitsch, N.J. Turner, C. Cassidy and D. Graham, Rapid and ultrasensitive determination of enzyme activities using SERRS, *Nat. Biotechnol.* 2004, **22**, 1133–1138.

7

Surface-Enhanced Infrared Spectroscopy

7.1 OVERVIEW

Soon after the identification of the surface-enhanced Raman scattering effect, the new field of surface-enhanced spectroscopy was unleashed, including linear [1,2] and nonlinear optical phenomena [nonlinear optical effects include the surface-enhanced second harmonic generation (SESHG), surface-enhanced double photon fluorescence (SEDPF), surface-enhanced hyper-Raman scattering (SEHRS), surface-enhanced four-wave mixing (SEFWM), surface-enhanced coherent anti-Stokes Raman scattering (SECARS) [3–5]. In particular, with Raman and infrared being the two faces of the vibrational coin, the search for the complementary surface-enhanced infrared absorption (SEIRA) was quickly rewarded with success [6]. Therefore, one can speak of surface-enhanced vibrational spectroscopy (SEVS), providing enhanced scattering and enhanced absorption techniques. The first SEIRA report by Hartstein *et al.* [6] was entitled 'Enhancement of the infrared absorption from molecular monolayers with thin metal overlayers', and was carried out using *p*-nitrobenzoic acid (PNBA) and recorded with attenuated total reflectance (ATR) in the infrared. The SEIRA spectra in the first report were only shown in the 2800–3100 cm^{-1} window. The enhanced infrared spectra reported were in fact due to C–H aliphatic stretching (CH_2 and CH_3, not PNBA) observed below 3000 cm^{-1}, as can be seen in Figure 7.1, drawn with spectra from Jensen *et al.* [7]. These characteristic CH_2 and

Surface-Enhanced Vibrational Spectroscopy R. Aroca

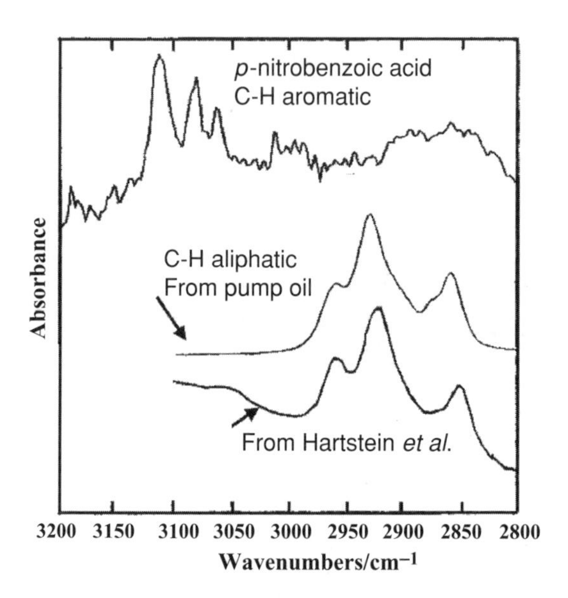

Figure 7.1 Infrared absorption spectrum of the aromatic C–H stretching vibrations of *p*-nitrobenzoic acid (PNBA) is shown at the top and the first surface-enhanced infrared absorption spectrum reported as of PNBA at the bottom with characteristic CH_2 and CH_3 stretching vibrations coming from the vacuum pump oil. After T.R. Jensen, R.P. Van Duyne, S.A. Johnson and V.A. Maroni, *Appl. Spectrosc.* 2000, **54**, 371–377 [7]

CH_3 stretching vibrations could have come from the vacuum pump oil employed in the evaporation system, also shown. The infrared absorption spectrum of the aromatic C–H stretching vibrations of PNBA deposited on a germanium internal reflection element are observed above 3000 cm^{-1}, as can be seen at the top of Figure 7.1 [7]. Therefore, the first enhancement was observed on the infrared spectrum of the impurity, not the intended target. It was clearly observed, however, that the intensity of the C–H aliphatic modes increased steadily with the thickness of the Ag overlayer, from 0 to 6 nm. At about the same time, the enhanced optical absorption in the visible region was the object of detailed studies, and a summary of the results up to 1981 can be found in the work of Eagen [8]. Eagen studied the enhanced absorption of a dye-coated silver island film, and showed that an absorption enhancement of the order of 10 could be obtained. In 1989, Muraki *et al.* [9] fabricated Langmuir–Blodgett (LB) monolayers of a dye (N,N'-dioctadecylrhodamine B) on silver island films and demonstrated the long-range character of the absorption enhancement in the visible region (wavelength from 420 to 750 nm), again with very modest enhancement factors. The largest enhancement in the absorption of the infrared radiation obtained from the metal overlayer work of Hartstein *et al.* [6] was a factor of 20. Clearly, the one-photon phenomenon cannot enjoy the large enhancement factor easily achieved in inelastic scattering, and the enhanced absorption may seem insignificant when compared with those of SERS. Nevertheless, it should be remembered that the average cross-section for infrared absorption is about nine orders of magnitude higher than the corresponding Raman cross-section, and a small enhancement in the infrared absorption can be extremely significant in terms of practical applications. However, some recent development in the near field may renew the interest in SEIRA using phonons rather than photon excitation [10]. To quote this reference:

> Here we study the strong enhancement of optical near-field coupling in the infrared by lattice vibrations (phonons) of polar dielectrics. We combine infrared spectroscopy with a near-field microscope that provides a confined field to probe the local interaction with a SiC sample. The phonon resonance occurs at 920 cm^{-1}. Within 20 cm^{-1} of the resonance, the near-field signal increases 200-fold; on resonance, the signal exceeds by 20 times the value obtained with a gold sample. We find that phonon-enhanced near-field coupling is extremely sensitive to chemical and structural composition of polar samples, permitting nanometer-scale analysis of semiconductors and minerals. The excellent physical and chemical stability of SiC in particular may allow the design of nanometer-scale optical circuits for high-temperature and high-power operation.

The phonon resonance is therefore the basis of SEIRA, where the near-field enhancement of the electric fields increases the infrared absorption of molecules coating the surface of certain nanoparticles. The analogy with the surface plasmons in metal nanoparticles is nicely summarized in Figure 7.2, a figure based on Figure 1 of Hillenbrand *et al.* [10], but

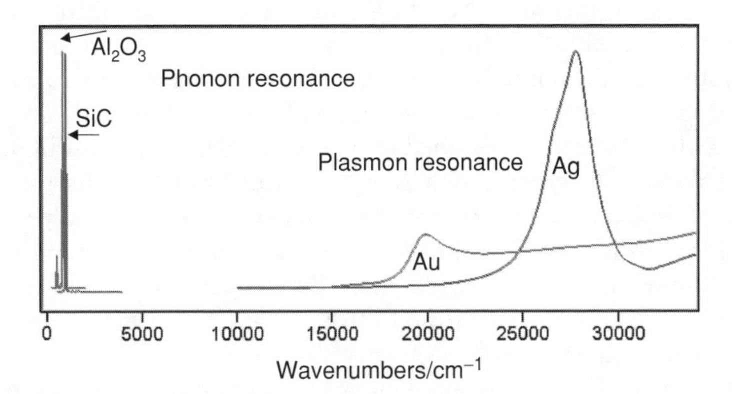

Figure 7.2 Phonon resonances of SiC and Al_2O_3 and surface plasmons in metal nanoparticles presented in the same spectrum

drawn with our own extinction Mie computations for metals and dielectric nanoparticles. In the figure, the locations of the resonances for a spherical particle of 20 nm radius of Ag, Au, SiC and Al_2O_3 are plotted in wavenumbers for an extended graph of the electromagnetic radiation from the visible to the mid-infrared region. The plasmon resonances of the two metals are seen in the visible region, whereas the phonon resonances of the dielectric particles are in the mid-infrared region below $1000\,cm^{-1}$. Since the intention is to illustrate the location of the resonances, the axis for the absorption intensity is arbitrary. The role of surface plasmons has been well documented (Chapters 2 and 3) and there is abundant literature on the subject, including a review of the plasmon literature directly related to SEIRA and SERS [11].

Soon after the seminal paper [10], Anderson [12] reported infrared enhanced absorption for anthracene coating polar dielectric nanoparticles of silicon carbide and aluminum oxide with 100-fold enhancement. The mechanism of SEIRA is explained as being the result of the enhanced optical fields at the surface of the particles when illuminated at the surface phonon resonance frequencies. This phonon resonance effect is analogous to plasmon resonance that is the basis for surface-enhanced absorption and enhanced Raman scattering in metals. The results obtained with dielectric nanoparticles open up new avenues for experimental SEIRA.

During the 1980s, SERS monopolized attention, and the initial sporadic activity on the subject of SEIRA was largely carried out in Japan, and a window into this early work can be found in Osawa's review [13]. In the 1990s, SEIRA received its share of attention, and there were several reports for both the practical and theoretical aspects of the phenomena [14–17]. Since the bulk of SEIRA work is recent, it is still of interest to demonstrate the effect itself, in particular the SEIRA spectra that can be obtained for the same system on different enhancing substrates. The effect has been observed on island films of the coinage metals and a few other surfaces, most notably Pt [17], Sn [18], Pd and Ru [19,20]. Looking for the plasmon in the infrared region led to the first study on metal films with architectures designed to produce surface plasmons in the infrared region [7], permitting a comparison of SEIRA results for these films and island films with strong plasmon in the visible and only a weak tail in the infrared region. SEIRA was observed in the region of the surface plasmon produced using engineered surfaces; however, when these results were compared with the SEIRA achieved on evaporated metal island films, equivalent enhancements were obtained.

The origin of SEIRA is attributed, by analogy with the interpretation accepted for SERS, to electromagnetic and chemical contributions responsible for the observed infrared enhancement. The analogy cannot be stretched too far, since the so-called chemical contributions to SERS include additional multiplicative effects, such as resonance Raman scattering [21,22] without analogy in absorption spectroscopy. The chemical effect in SEIRA is a generic term used to indicate several variations observed in the infrared vibrational fundamental. They are the result of molecule-enhancer interactions that may affect the frequency, the shape of the observed infrared band or the intensity of fundamental vibrational mode that is determined by the partial derivative $(\partial\mu/\partial Q)_0$. In addition, the adsorbed target molecule may display a different symmetry than that of the free molecule, as was discussed in Chapter 4 for SERS.

The electromagnetic (EM) contribution is attributed to enhanced optical fields and, correspondingly, augmented absorption intensities are plasmon or phonon assisted. In his broad review on SEIRA, Osawa [13] discusses the plasmon-assisted models, which can quantitatively explain the effect, and the applications to the study of electrochemical reactions. Here, we closely follow our own review on the subject [23]. When the surface underneath the adsorbed monolayer is a smooth reflector, the infrared spectroscopy is the reflection–absorption infrared spectroscopy (RAIRS) mode, which is well explained in terms of Fresnel's equations and optical properties of the metal surface and strictly obeying the surface selection rules. Both SEIRA and SERS found their EM origin in light absorption by nanoparticles. The explanation of the origin by plasmon assistance has been put in plain words in many reports, as for example [24]

> According to classical electrodynamics and in agreement with Mie's theory, one feasible mechanism of this effect is as follows. The electrons inside of nanoparticles follow the applied E field and transmit dipole radiation. The transmitted E field owns a frequency and wavelength similar to the applied E field. Within certain size ranges of the nanoparticles, there is a resonance and certain phase relation between the electrons in the nanoparticles and the light wave. This resonance and phase relation is supposed to cause the effect of enhanced absorption. In other words, the nanoparticles act as little antennas which are just tuned right for a high interaction with the light wave.

Eagen [8], describes the enhanced absorption in the visible as being "...due to the amplification by the Ag core of the electric fields associated with the driven modes of the dye layer". In summary, in the electric dipole approximation, the coupling operator $\hat{H} = -\mu \times E$ (see Chapter 1) for a vibrational transition is enhanced due to a local field E_{loc} that is much larger than the incident field E. In classical terms, for molecules with a fixed spatial orientation, the active infrared modes of a given symmetry species will be seen with an absorption intensity proportional to $(E \cdot \mu_i)^2$, the square of the scalar product where μ_i is the component of the dynamic dipole in the direction i of the optical field E. Since the E_{loc}/E, the local field to incident field ratio, can be >1 (usually between 1 and 30), the corresponding enhanced absorption would be directly proportional to the square of field enhancement.

Experimentally, the enhanced optical field can be achieved using substrates with discrete and non-discrete surface plasmons. An example of the first type are the ordered arrays of uniformly sized silver nanoparticles, prepared by nanosphere lithography (NSL) [7]. The non-discrete surface plasmon substrates are the commonly used Ag and Au island films that are tuned towards the infrared by increasing the mass thickness of the metal island film. For instance, Osawa and Ikeda [25] showed that SEIRA of *p*-nitrobenzoic acid deposited on Ag island films, evaporated on to a CaF_2 substrate and with no discrete surface plasmon in the infrared, increases as the mass thickness of the metal film increases from 4 to 10 nm, and the trend is reversed after a 14 nm mass thickness has been deposited. The most distinct EM property, the distance dependence of SEIRA, has been demonstrated using LB monolayers. A varying number of arachidic acid spacer layers were deposited on silver island films before deposition of the probe monomolecular layer containing functional groups (carbonyl) with a large infrared absorption cross-section. Similarly to the results found for the enhanced absorption in the visible region of the EM spectrum, SEIRA is most efficient within 5 nm of the surface.

EM local field enhancement can be predicted using, for instance, effective medium theories [17,26] and there is no question that there is a key contribution from enhanced local fields to SEIRA. There is an increase in the rate of absorption per unit volume that is proportional to the energy density of the field at the appropriate frequency. The enhanced local field augments this energy density at the surface of particles where the adsorbed molecule resides. This local field varies according to several factors, including size, shape and the dielectric function [27]. The enhancement varies from point to point and the average value is expected to match the observations. There are further consequences for the observed infrared spectrum from molecules adsorbed at these local fields. The local field may be highly polarized. Moskovits [2] has illustrated the implications of having a perpendicular polarized field, or a tangentially polarized field, in determining the surface selection rules. For metals, in the infrared region, a perpendicular polarized field can be assumed to be predominant. As pointed out before, the metal interacting with the adsorbed molecule (chemical effect) may induce changes in a fashion similar to what is seen in electrochemistry [28], producing a variation in the dipole moment derivatives and hence in the infrared intensity.

Bjerke *et al.* [17] reported a peculiar property in the symmetry of the SEIRA band shape of CO on platinum. The band asymmetry was further investigated and was also observed on Ag and Au island films. This new damping effect was simulated and explained using an effective medium approach. Krauth *et al.* [29] measured, in ultrahigh vacuum (UHV), the infrared transmission spectra of CO on ultrathin films of Fe grown at about 315 K on UHV-cleaved MgO(001) and observed enhanced asymmetric CO stretching bands. Later, the same group showed a correlation of the asymmetry of the enhanced CO stretching line and the electronic properties of the underlying ultrathin epitaxial iron metal film [30]. This 'Fano band' shape attracted attention owing to a possible link of the observations to the dynamic interaction of the adsorbate vibrations with electron–hole pair excitations [31]. Expressions for the lineshape of an isolated vibrational mode in the presence of electron–hole damping have been derived by Landgreth [32]. It was shown that the electron–hole decay mechanism produces an asymmetric line shape. Priebe *et al.* [33] also reported the surface-enhanced infrared absorption of CO on iron films. Notably, in a detailed study published in 2003, the group established the correlation between the enhancement and the asymmetry of the CO band with the surface plasmon [34]. They summarized their findings as follows:

> Various experimental results on surface enhanced infrared absorption reveal asymmetric line shapes. Whereas the order of magnitude of the enhancement can be understood from electromagnetic field enhancement the unusual line shape remains without satisfactory explanation. An interaction with electron–hole pairs would lead to an asymmetric line but this should be restricted to the first monolayer. However, asymmetry is also observed for vibrations at larger distances from the metal–film surface. Here we show strongly asymmetric lines and their enhancement as a consequence of the interaction of adsorbate vibrations with surface plasmons of metal islands. Both the effects and also the baseline change can be estimated by a proper application of well established effective-media models.

In summary, the enhancement factors (EF) observed in SEIRA are usually found to be in the $1 < EF < 100$ region. Phonon-assisted SEIRA may produce higher values than those obtained on metal nanoparticles. The local field enhancement of electromagnetic models can account for the enhancement and the asymmetry of bands observed with small molecules. The molecule-enhancer interactions could leave a very well-defined signature in the observed IR spectrum (chemical effect) expressing the distinct vibrational properties of a surface complex with its own symmetry, force field and chemical composition. In addition, with or without the formation of a surface complex, the molecular orientation and polarization of the local field will determine what is observed, with the caveat of the depolarization and extra scattering that are a ubiquitous part of an enhancer made of nanoparticles and aggregates of nanoparticles.

The SEIRA enhancement can be used to detect monolayers and submonolayer coatings on enhancing metal films [15] and, although SEIRA does not yield the enhancement necessary for single-molecule detection as SERS does [35,36], SEIRA is a viable means of enhancing the infrared signal from adsorbed molecules on a variety of metals, semimetals, semiconductors and polar dielectric nanostructures.

7.2 THEORETICAL MODELS FOR SEIRA

The optical properties and the electromagnetic field enhancement on rough metal surfaces and nanoparticles have been extensively discussed for radiation in both the visible and near-infrared regions of the spectrum [37–39] (see Chapter 3 and references therein), and has also been extended to the mid-infrared region [23,26,40]. Computational approaches for SERS electromagnetic enhancement [41–43] are often applicable to SEIRA, and they commonly model isolated particles, aggregates of nanoparticles or collection of particles in thin films. The single-particle models and aggregates were at the center of the discussion of SERS, while the effective medium approach suitable for the treatment of a collection of particles in thin films has been used for SEIRA.

In the seminal paper on 'Colours in metal glasses and in metallic films', Maxwell-Garnett [44] discussed color in metal glasses containing a small amount of minute metal spheres and that is followed by a second model where metal films are treated allowing the metal proportion to vary from zero to unity. In this paper it is proven that a medium formed by small metal spheres (much smaller than the wavelength of light) is optically equivalent to a medium with an effective dielectric value given by [37]

$$\bar{\varepsilon} = \varepsilon_h \frac{3 + 2\sum_i f_i \alpha_i}{3 - \sum_i f_i \alpha_i} \tag{7.1}$$

where α is the polarizability of the inclusions, i is the index over different particles, $\bar{\varepsilon}$ is the effective dielectric function, ε_h is the dielectric function of the host material, in which the particles are embedded, and f represents the volume fraction of the inclusions. This is the fundamental principle of the effective medium theories that can allow for the distribution of particle shapes and sizes to be included in the calculation, in addition to simple spheres.

The most common substrates used in SEIRA are metal island films, or granular materials (such as metal colloids). The purpose of the effective medium theory (EMT), or effective medium approximation (EMA), is to simulate the dielectric functions of the microscopically inhomogeneous nucleating layers or discontinuous films used in the models [14,15,26,45]. In the case of SEIRA, the effective optical property represents an average for the metal films, the substrate and the organic coating. Therefore, the electromagnetic SEIRA enhancement may be calculated using effective medium theories, which involves finding the effective optical properties of the mixture of these components. One of the most often used formalisms for effective medium calculations is the Bruggeman method [46,47] A very general review of the physical properties of macroscopically inhomogeneous media, and the applications of EMT to electrical and electromagnetic properties, can be found in the extensive and excellent review of Bergman and Stroud [48].

The Bruggeman model is a self-consistent theory which includes a greater amount of interaction between inclusions. The self-consistency enters through the use of the Bruggeman condition, $\varepsilon_h \rightarrow \bar{\varepsilon}$, which requires the solution to be a dielectric function of a host that has the same optical properties as the effective medium. This formalism can be expressed in general as

$$\bar{\varepsilon} = \varepsilon_h \frac{3(1 - f) + f\alpha'}{3(1 - f) - 2f\alpha'} \tag{7.2}$$

where $\alpha'$is the polarizability of the inclusions with the 'Bruggeman' condition applied.

There are several reports where EMTs have been used to model SEIRA experiments. Osawa *et al.* [45] modeled island films as a set of ellipsoids covered with a dielectric film, where the symmetry axis is normal to the substrate. Bjerke *et al.* [17] used the Bergman's representation of the effective dielectric function to model their observations of enhanced absorption of the infrared spectrum of adsorbed CO on platinized Pt surfaces. By varying the platinization conditions, platinized Pt surfaces yielded SEIRA enhancements of up to 20 times that of CO adsorbed on smooth Pt electrodes. Using the Bergman EMA, they also provided a simulation for increase CO band asymmetry observed as platinization was increased. The unique feature in Bergman's approach is the attempt to incorporate the microgeometry in the calculation of the bulk effective dielectric constant [49,50].

In practical calculations, the polarizability of a coated ellipsoid is used, with depolarization factors derived in 1945 by Stoner [51] and Osborn [52]. Since the inhomogeneities in the layer are much smaller than the wavelength of the incident light, it is assumed the mixed film is a continuous, parallel-sided layer, so Fresnel's equations may be used to calculate the reflectance and transmittance [53]. Therefore, discontinuous metal films consist of islands, which are modeled as ellipsoids of revolution or spheroids of uniform shape and size [46]. There are two different types of ellipsoids of revolution: oblate, where the two larger axes are equal, and prolate, where the two smaller axes are equal. It should be pointed out that Normann *et al.* [54] showed that prolate spheroids with the rotation axis parallel to the plane provide the best description of electron micrographs and also the best fit between measured and computed spectra. The characteristic plasmon resonances for metals in this region lead to strong absorption bands for most noble metal, and these spectral features are strongly dependent on the shape and size of the metal particles. The first approximation to the geometry for the metal islands can be found by fitting computed UV–visible plasmon spectra to measured data in the same region. Flat islands have a major axis of length a, minor axis b and an aspect ratio defined as $\eta = a/b$.

The dielectric function values, denoted ε_m, used in the computations are taken from those for the bulk metals collected in the *Handbook of Optical Constants of Solids* [55]. Notably, Roseler and Korte [56] have determined the optical constant of metallic island films and reported that they differ from the bulk values used for surface enhanced infrared absorption calculations.

Additional considerations of the inhomogeniety of the size and shape of the particles can be taken into account by using a distribution. Since real films are not made up of particles of only one size and aspect ratio, a distribution of the axis of particles would be a better model for the film [54]. It has been found that a log-normal distribution provides the best fit for electron microscopy data [57]. The log-normal distribution is given by

$$f_{\mathrm{LN}}(x_s) = \frac{1}{(2\pi)^{\frac{1}{2}} \ln \sigma_s} \exp\left\{ -\frac{1}{2} \left[\frac{\ln\left(\frac{x_s}{\bar{x}_s}\right)}{\ln \sigma_s} \right]^2 \right\} \qquad (7.3)$$

where σ_s is the standard deviation of the length of the minor axis, x_s the length of the minor axis and $\overline{x}_s$ the average length of the minor axis. In general, a larger deviation will result in a broader plasmon, since contributions from many different shapes will form the spectra.

The introduction of the organic layer to the metal surface can be represented in two ways. The first is to assume that the molecule forms a thin layer uniformly coating the surface. The net dipole moment p of the layered spheroid can be written as $p = \alpha V E_{\mathrm{loc}}$, where α is the polarizability, V is the volume of the inclusion and E_{loc} is the local field, made up of the incident field and the interaction fields from the particle. For a coated spheroid with one axis parallel to the incident electric field, the polarizability is given by [8,58]

$$\alpha = \sum_i \frac{1}{2} \frac{(\varepsilon_d - \varepsilon_h)\left[\varepsilon_m L_{1i} + \varepsilon_d (1 - L_{1i})\right] + Q(\varepsilon_m - \varepsilon_d)\left[\varepsilon_d (1 - L_{2i}) + \varepsilon_h L_{2i}\right]}{\left[\varepsilon_d L_{1i} + \varepsilon_h (1 - L_{2i})\right]\left[\varepsilon_m L_{1i} + \varepsilon_d (1 - L_{1i})\right] + Q(\varepsilon_m - \varepsilon_d)(\varepsilon_d - \varepsilon_h) L_{2i} (1 - L_{2i})} \tag{7.4}$$

where ε_m is the dielectric function of the metal, ε_d the dielectric function of the organic layer, ε_h the dielectric function of the host medium, Q the volume ratio of the ellipsoid, defined by $Q = V_{\mathrm{core}}/V_{\mathrm{coat}}$, i the index over the axes of the inclusion and L_{1i} and L_{2i} the geometric factors corresponding to the core ellipsoid and the coated ellipsoid, respectively.

Basic to the definition of ellipsoids is the geometric factor. The geometrical factor is a measure of the curvature perpendicular to a specific axis of the ellipsoid, and has a value $0 < L < 1$. The geometric factors for the major axis of an oblate are [46]

$$L_1 = \frac{g(e)}{2e^2}\left[\frac{\pi}{2} - \tan^{-1} g(e)\right] - \frac{g^2(e)}{2}, \quad g(e) = \left(\frac{1 - e^2}{e^2}\right)^{\frac{1}{2}} \tag{7.5}$$

and for the major axis of a prolate

$$L_1 = \frac{1 - e^2}{e^2}\left[-1 + \frac{1}{2e} \ln\left(\frac{1 + e}{1 - e}\right)\right] \tag{7.6}$$

where e is the eccentricity of the spheroid.

The second method of introducing organic molecules to the metal surface is to model the inclusions directly embedded in the organic matrix. This model is better suited to thicker layers of organic material, since a coating around a particle poorly describes this case. In this model the polarizability is given by

$$\alpha = \sum_i \frac{1}{2} \frac{\varepsilon_m - \varepsilon_d}{\varepsilon_d + L_i(\varepsilon_m - \varepsilon_d)} \tag{7.7}$$

with the same notation as above, except that only one geometric factor is needed.

The Bruggeman EMT tends to give higher enhancements, and has often been used to explain some of the larger experimentally observed enhancements. However, these experiments also used molecules which chemisorbed, and so some of the observed enhancement may be due to the chemisorption of the molecule on to the surface [25].

Effective medium theory is popular model for SEIRA as enhancement factors can be calculated without any difficulties regarding the nature of the enhancement. Although useful, effective medium theory does not give any insight as to the mechanism of the phenomenon.

Typical EM results are illustrated here first for field enhancement of isolated metal particles based on the approach of Zeman and Schatz [59] used to study enhancement factors for a variety of materials in the visible region. The method can be extended to the full infrared region and the result is similar to the static case due to the long wavelengths. Figure 7.3 shows the average field enhancement calculated for an isolated spheroid of Pt, Ag, Au and Sn in vacuum with lengths of 90 and 30 nm for major and minor axes, respectively.

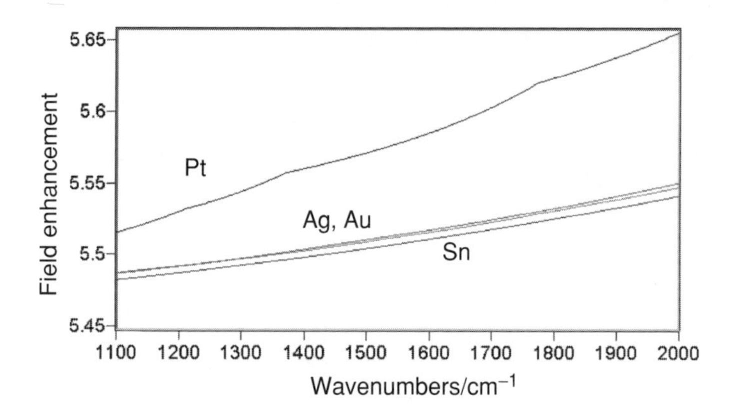

Figure 7.3 Calculations showing the average field enhancement in vacuum for isolated spheroids of Pt, Ag, Au and Sn, with major axes of 90 nm and minor axes of 30 nm

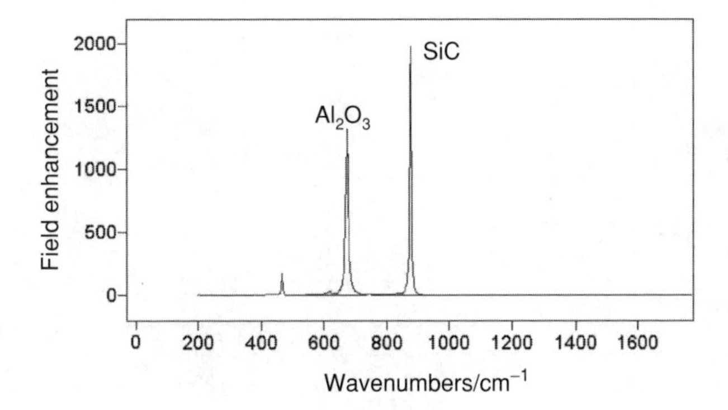

Figure 7.4 Enhancement field calculations, averaged over the surface of the particle, for isolated spheroids of SiC and Al_2O_3

Enhancement field computations for isolated spheroids of SiC and Al_2O_3, 90 nm long by 30 nm wide, averaged over the surface of a particle are shown in Figure 7.4.

Examples of effective medium calculations are presented in Figures 7.5 and 7.6, showing the Maxwell–Garnet (MG) and Bruggeman calculations, respectively, for a collection of prolate ellipsoids, with a major axis of 40 nm and a minor axis of 20 nm and uniformly coated with a 1 nm thick layer of the organic material. The organic coating in these model calculations is 3,4,9,10-perylenetetracarboxylic-dianhydride (PTCDA), a well-known organic dye [60]. The dielectric function for the organic is

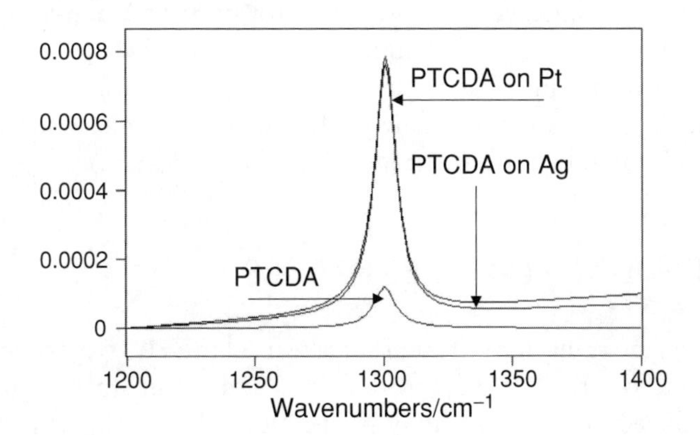

Figure 7.5 Effective medium calculation results using the Maxwell–Garnet approach for a collection of prolate ellipsoids uniformly coated with a 1 nm thick layer of the organic material

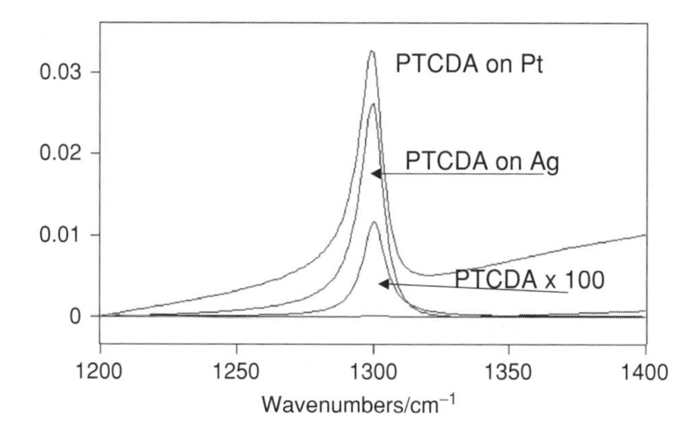

Figure 7.6 Effective medium calculation results using the Bruggeman approach for a collection of prolate ellipsoids uniformly coated with a 1 nm thick layer of the organic material

taken from the PTCDA infrared spectrum using the absorption band at 1300 cm^{-1}, using a Lorentz model with an FWHM of 10 cm^{-1}, giving the effect of the metal.

The Bruggeman EMA results produced larger enhancement factors than the MG approximation, as can be seen by comparison of the two figures. Some experimental results seem to agree with Brugeman predictions of higher SEIRA enhancement factors ($\sim 10^2$) than those predicted by the MG approach. For instance, early on, Osawa and Ikeda [25] reported enhancement factors of ca 500–600 for *p*-nitrobenzoic acid on silver island films. Very recently, spectra of SEIRA-active substrates fabricated by glancing angle vapor deposition (GLAD) [61], a thermal evaporation technique that produces aligned Ag nanorods, have been compared with the corresponding RAIRS spectra of the same adsorbate with an estimated enhancement factor of about 53 [62].

7.3 SEIRA-ACTIVE SUBSTRATES

Commonly, the same active rough surface used in SERS experiments may also be used for SEIRA. Apart from metals and semimetals, semiconductors and dielectric materials may be used to enhance the absorption in the infrared region of the spectrum. However, since SEIRA enhancement depends on the size, shape and particle density of the selected metal island films, fine tuning of these parameters to maximize the SEIRA signal is necessary. For the IR experiment, the SEIRA-active nanostructures are fabricated on a supporting substrate that must be an infrared-transparent material for transmission geometry, or a reflecting substrate for the surface-enhanced reflection absorption (SEIRRA) experiment [15]. The most commonly used experimental configuration is the metal–underlayer, where the sample is deposited onto the active nanostructure fabricated on an infrared-transparent substrate. The supporting substrates, commonly available, include infrared-transparent materials such as Si, Ge, CaF_2, BaF_2, KBr, ZnSe, ZnS, KRS-5, sapphire and MgO used in the transmission experiment, and non-transparent substrates such as glass, glassy carbon, polymers and metals are only adequate for external reflection measurements. When the SEIRA-active material is deposited by thermal evaporation, the film morphology is influenced by the surface structure of the supporting substrate, and also by the experimental conditions used during the film deposition (evaporation rate and temperature of the supporting substrate). Metal–overlayer (metal–sample) and even sandwich (metal–sample–metal) [63] configurations have also been used successfully. Notably, the dielectric properties of the substrate (refractive index) may play a substantial role in the observed spectra, particularly in the SEIRRA experiment, where the observation of positive and negative absorptions depends on the angle of incidence and the polarization of the incident radiation. Distortion of the band shape in the corresponding SEIRA/SEIRRA spectra may also be seen.

The SEIRA-active metal islands are commonly prepared by high-vacuum evaporation of the metal on to a supporting substrate [15,64]. The typical system includes a stainless-steel chamber, a glass bell-jar, a tungsten basket for placing the metal, a quartz crystal microbalance (QCM) for monitoring the film thickness and the required vacuum pumps. Low-melting points metals are evaporated by resistive thermal heating of the tungsten boat, while direct resistive heating of metal wires is used for metals with high melting points. Alloys can be prepared through simultaneous evaporation of both metals. The deposition rate of the metal is a crucial parameter for the shape and size of the islands, with slow rates (0.1–0.5 nm min^{-1}) generally giving the best enhancement. The optimal film thickness for maximum enhancement depends on the deposition rate; in fact, the optimization of both parameters must be made for each metal–substrate system. In an effort to control the morphology of the evaporated film and, therefore, its corresponding surface plasmon region, templates such as periodic particle array (PPA) films prepared by nanosphere lithography (NSL) have been employed [7]. Anodized porous Al surfaces and island films of different metals have been used as templates for subsequent metal evaporation.

The equipment for metal vacuum evaporation is routinely used in surface laboratories; however, it is not readily available in most infrared laboratories. An alternative, and less expensive, method to form metal islands for SEIRA is electrochemical deposition, where a suitable potential or current is applied to electrolyte solutions of salts of the metal to be deposited, under either potentiostatic, galvanostatic or potential cycling conditions. The surface roughness, or island size, may be controlled by changing the concentration of the solution, the voltage or current applied and the time utilized in the deposition. In some cases additives are used to control the morphology. The thickness of the film can be estimated from the charge passed during the deposition. In this method, the substrate on which the metal is to be deposited must have high electrical conductivity. For this reason, materials such as glassy carbon or bulk metal have usually been used, although metal electrodeposition on Au-coated glass and n-type Si has also been reported. Electrochemical deposition has mainly been employed to prepare rough electrodes used in spectroelectrochemical studies, but its use has been proposed for manufacturing large amounts of recyclable SEIRA surfaces efficiently and at low cost [65]. The electrochemical preparation of gold particles and rods in anodic alumina templates has also been reported [66].

Metal colloids can be employed as SEIRA-active nanostructures. They are prepared by reduction of metal ions. Although aqueous metal colloids are one of the most common media employed in SERS due to the very low Raman scattering cross-sections of water, the extremely high infrared absorption of water prevents their straightforward application in infrared measurements. Nevertheless, SEIRA enhancement using silver and gold colloidal particles has been reported. For instance, once the aqueous silver colloid has been prepared, one aliquot amount (around 500 μL) is cast on to a KRS-5 substrate and dried out. The process is repeated to obtain thick aggregated colloidal films [67]. On the other hand, colloidal gold bioconjugates (gold–protein complexes) have proved to be useful in immunoassays based on the SEIRA effect, either as wet samples collected by filtration on a porous polyethylene membrane or as dry films [68]. Another possibility of using colloidal particles as SEIRA-active surfaces is to immobilize them on silane-derivatized glass substrates. External reflection SEIRRA spectra have been reported for 1,5-dimethylcytosine (1,5–DMC) [69] using, as active surfaces, laser-ablated silver colloids that were immobilized on an AMPTS (3-aminopropyltrimethoxysilane)-derivatized glass slide, previously reported in SERS experiments [70]. The SEIRA spectra of 1,5-DMC on 'evaporated' and 'immobilized' colloidal Ag nanoparticles are equivalent.

Metal sputtering (magnetron, diode) has also been used for the deposition of nanoparticles of Ag [16], Au [71] and Pt [72] for SEIRA experiments.

Recently, the chemical (or electroless) deposition technique, where the substrate is immersed in suitable metal plating solutions, has been applied successfully in preparing nanoparticle films of Au [73], Ag [74,75], Pt [76] and Cu [77]. AFM images of the chemically deposited Au films revealed an island structure similar to that of the vacuum-evaporated Au films, with larger average dimensions of the islands (300 instead of 70 nm). The chemical deposition technique seems to be simple and cost-effective with good SEIRA enhancement. The technique also provides better adhesion of the metal nanoparticles to the substrate and less contamination (in the case of vacuum-evaporated films, hydrocarbon contaminants existing between the metal and the substrate may interfere the spectral measurements, unless the surface is first plasma treated). The adhesion is a relevant property when the deposited island film is used in an electrochemical environment. SEIRA has been extensively used in electrochemistry research with metal films deposited on the silicon prism employed for *in situ* measurements with the Kretschmann ATR configuration [78,79]. Particularly, evaporated gold films adhere very poorly to silicon and the standard adhesion promoters have a detrimental effect on their island structure and SEIRA activity. As a result of the efforts made to find new methods for the preparation of island Au films with suitable adhesion and enhancement factors, new recipes have recently been published. One of them employs chemical deposition with HF etcher added to the plating solution [73] and the other is based on etching by aqua regia of a thin Au film initially deposited on the Si prism by thermal evaporation [80]. In both cases, etching of the silicon surface either with HF or with NH_4F solutions, to remove oxide layers and to terminate it with hydrogen, was carried out. This pretreatment provides enhanced adhesion because it facilitates the formation of a silicide between Si and Au deposits.

Modified SEIRA-active surfaces based on coating the metal substrate with self-assembled monolayers (SAMs) of different thiols have been tested and evaluated, both for gold–silicon [81,82] and for gold–germanium systems [83]. The goal is to extend the applications of the SEIRA technique, making it possible to achieve some SEIRA enhancement in the case of molecules that, in principle, do not show SEIRA spectra. Such could be the case of molecules without polar groups that, consequently, have no possibility of being chemisorbed to the metallic surface. If the 'modified substrates' succeed in placing the molecules close enough to the metallic particles for sensing the enhanced electromagnetic field, their infrared spectra would benefit from such enhancement. In particular, sulfur and selenium compounds have a strong affinity to transition metal surfaces and, specifically, SAMs of thiols on gold and silver have been the subject of considerable research because such molecules spontaneously chemisorb to form densely packed and structurally ordered thin films. This facilitates the process of obtaining such SAMs modified SEIRA substrates.

The coinage metals Ag and Au, which show high enhancement factors in SERS, have also been the most widely employed metals in SEIRA measurements. However, SEIRA models predict enhancement on transition metals as strongly as on coinage metals [13]. In practice, SEIRA has been observed on many other metals, e.g. Cu [84], Sn [18], Pb [85], Fe [33], Pt [17,78,79], Ni [86], Pd [86], Rh [87], Pt/Ru [20], Ir [88] and Pt–Fe alloys [89], with varying magnitude of enhancement and with some peculiarities affecting the SEIRA band shapes. Ag and Au island films are the most commonly used SEIRA substrates, but in electrochemical experiments the platinum group metals are preferred. In particular, Pt, Pd and Rh have been extensively employed as electrode materials for electrochemistry, and because they are commonly used as heterogeneous catalysts in a wide variety of industrial processes, their SEIRA enhancer possibilities have been actively explored [17,19,78,79,90]. The investigation and explanation through adequate SEIRA models of the abnormal infrared effects (AIREs) observed in the bands of the SEIRA spectra of molecules adsorbed on these metals and on Fe [34,91] are ongoing. The extension of the SEIRA substrates to dielectric materials is a more recent development [12]. Enhanced absorption of the vibrational spectrum in the infrared region has also been reported on planar Ag halide fibers, showing 10-fold spectral amplification that could be due to the increased coupling of evanescent waves through an increased number of internal reflections [92,93]. However, the authors have classified their observations under the umbrella of SEIRA.

Several ways of placing the sample under study on the SEIRA-active substrates have been reported. The easiest approach is the 'cast film method' or 'drop-drying method', where films of the sample are cast by dropping their dilute solution with a microsyringe over the metallic surface. The solvent is allowed to evaporate slowly, leaving an ultrathin film of the sample on the active surface. A thin film of the sample can also be formed by evaporation in vacuum on the metal surface, or formed as an LB film using a thin-film deposition device attached to a Langmuir balance for transfering to the SEIRA-active substrate [15].

When the sample under study is known to chemisorb on the metal of the SEIRA substrate, the 'SAM method' can be applied. In this case, the SEIRA-active substrate is immersed in a solution of the sample and allowed to soak for some period. Subsequently, the substrate is rinsed thoroughly with the solvent so that a SAM of the sample remains on the metal island surface. In electrochemical environments, the sample is in solution where the roughened electrode (SEIRA-active surface) is used to probe the species adsorbed at the interface.

Transmission FTIR and internal and external reflection are now commonly used in SEIRA measurements. On occasion, the diffuse reflection mode has been used for dispersed metallic colloids. However, its use to increase the SEIRA sensitivity for species on the surface of strongly absorbing materials, such as carbon-based catalysts and natural vegetation, is under investigation. For electrochemical studies, ATR–SEIRA with the Kretschmann configuration (prism–thin metal film–solution) is the best option [64].

7.4 INTERPRETATION OF THE OBSERVED SEIRA SPECTRA

Electric dipole selection rules are key to determining whether absorption of electromagnetic radiation by an oscillating dipole is possible [94]. Therefore, the absorption of light by an oscillating molecule in the ground electronic state is given by the coupling of the dielectric dipole to an external electromagnetic field: $H' = -\mu \cdot E$ [95], and the probability of absorption is thus proportional to the square of the dipole moment matrix element along the direction E_j of light polarization (see Chapter 1). The amplitude of the transition is proportional to the scalar product $j \cdot \langle \psi_0 |\mu| \psi_1 \rangle$, where the wavefunctions are those of the harmonic oscillator [96,97]. In the gas phase, where all molecular orientations have equal probability, the geometric part of the scalar product is a constant. Therefore, the selection rules for fundamental transitions between vibrational levels ψ_0 and ψ_1 are determined by the matrix element $\langle \psi_0 |\mu| \psi_1 \rangle$. Since we are discussing the ground electronic state, the dipole moment μ is a function of the vibrational coordinates and for infinitesimal vibrations it can be written in a Taylor series as described in Chapter 1. In Cartesian coordinates, the dipole moment derivative is a vector, $\boldsymbol{\mu}' = \boldsymbol{\mu}_x{}' + \boldsymbol{\mu}_y{}' + \boldsymbol{\mu}_z{}'$ with components $\boldsymbol{\mu}_i{}'$. In the ground electronic state, the equilibrium geometry determines the symmetry point group of the molecule, simplifying the application of selection rules by including the vector components different from zero in the character table of each point group. Symmetry reduces the electric dipole selection rules for an allowed transition between two vibrational states, connected by an operator, to the requirement that the direct product (triple product) has a totally symmetric component (reference, 98 p. 128). For an isolated molecule or gas-phase spectra, the triple product is directly given in the character table.

In the solid state or molecules adsorbed on surfaces with a fixed spatial orientation, the intensity of a vibrational transition would be proportional to $(E \cdot \mu_i')^2$, where μ_i' is the component of the dynamic dipole in the direction i of the electric field $\boldsymbol{E}$. Therefore, for allowed infrared modes of a given symmetry species to be seen with maximum vibrational infrared absorption intensity, there must be alignment between the polarization of electric field vector $\boldsymbol{E}$ and one of the non-zero components of the dynamic dipole moment derivative. On reflecting metal surfaces, the incident and the reflected electric field vectors form the plane of incidence. Electromagnetic waves with the electric field normal to the plane of incidence are denoted as s-polarized, transverse electric (TE) or $\perp$ waves. When the electric field is parallel to the plane of incidence a wave is called p-polarized or transverse magnetic (TM). In 1966, Greenler [99], studying adsorbed molecules on metal surfaces by reflection techniques, realized that at high angles of incidence, p-polarized light "has a sizable component of the electric vector normal to the metal surface". The opposite was true for the s-polarized light. Therefore, for molecules adsorbed on metal surfaces, the maximum vibrational intensity is derived from $(E_p \cdot \mu_i)^2$. The latter is the simplest expression of the 'surface selection rules' discussed in Chapters 2 and 3 with an example given for the reflection–absorption infrared absorption. The reduction of the surface selection rules for metal surfaces to the statement 'that only the vibrational modes having nonzero dipole moment derivative components perpendicular to the surface were infrared active' was also advanced by Pearce and Sheppard [100] and Hexter and Albrecht [101], explaining the observation in terms of the induced image dipoles (Chapter 2). A dipole moment change parallel to the surface is cancelled by a dipole moment change of the same magnitude in the opposite direction, induced in the substrate, while dipole moments changes perpendicular to the surface are reinforced and the total dipole moment would be doubled.

3,4,9,10-Perylenetetracarboxylic dianhydride (PTCDA) dye is one of the archetypal flat molecules producing ordered films over extended distances, which has been grown on various substrates such as highly oriented pyrolytic graphite (HOPG), MoS_2, Ag(111), Ag(110), Ni(111), Ge(100), Cu(100), Au(111) and GaAs(100) [102]. It provides an ideal

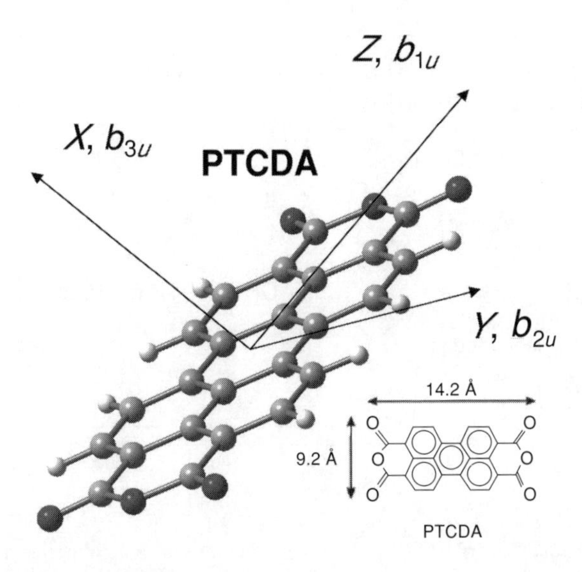

Figure 7.7 Structure and symmetry species allowed in the infrared spectrum of PTCDA

study case to illustrate the application of the selection rules in the spectral interpretation of the FTIR transmission, reflection–absorption infrared spectroscopy (RAIRS) and SEIRA [23]. The molecule has 38 atoms ($C_{24}H_8O_6$), 108 vibrational degrees of freedom and 46 infrared-active fundamental vibrations [103] distributed as $10b_{3u}(x) + 18b_{2u}(y) + 18b_{1u}(z)$ symmetry species. The molecule is a flat rectangle of 14.2 Å for the long axis (z) and 9.2 Å for the short axis (y), as shown in Figure 7.7.

Since the molecule has a center of symmetry (D_{2h} group), the mutual exclusion rule applies. RAIRS is an infrared technique that takes advantage of the large E_p component at the reflecting surface. In the present study case, the samples for RAIRS were prepared by evaporating first 100 nm of Ag on to a Corning 7059 glass slide, held at 200°C under high vacuum. A 50 nm layer of PTCDA was then evaporated on to this smooth silver surface [104]. The RAIRS spectra were obtained by using a Spectra-Tech variable-angle reflectance accessory set such that the incident beam impacted the surface at 80° from the normal.

The reference is the transmission spectrum of the solid dispersed in a KBr pellet that is equivalent to a random distribution of the PTCDA molecules, i.e. a random molecular orientation. To help the vibrational analysis, a calculation of the vibrational frequencies and intensities is carried out. To illustrate the agreement between the experiment and DFT computations, the transmission FTIR spectra of a PTCDA pellet and

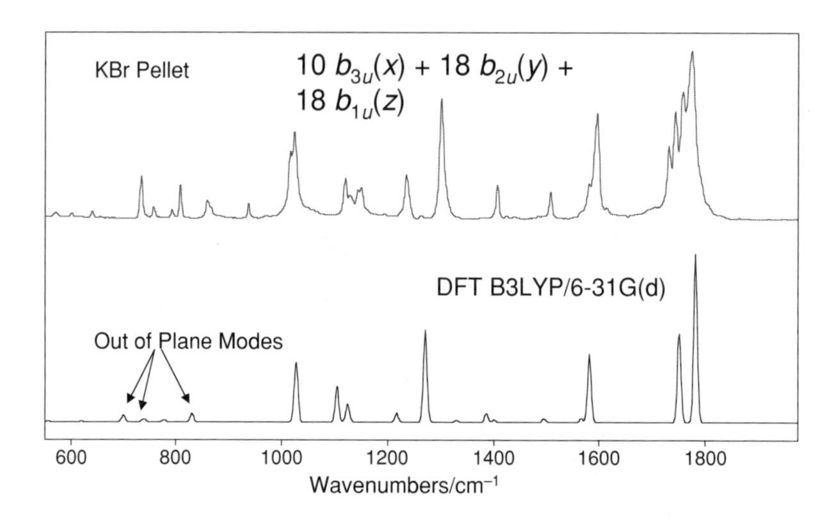

Figure 7.8 Transmission FTIR spectrum of PTCDA in a KBr pellet and *ab initio*-calculated infrared spectrum using DFT B3LYP/6–31G(d). Reproduced from R.F. Aroca, D.J. Ross and C. Domingo, Surface-enhanced infrared spectroscopy, *Appl. Spectrosc.* 2004, 58, 324A–338A by permission of Applied Spectroscopy

the B3LYP/6–31G(d) results obtained using Gaussian 98 [105], where a scaling factor of 0.9614 [106] was used, are shown in Figure 7.8. It can be seen that there is good agreement between the calculated and observed vibrational intensities. The latter is in spite of the condensed matter effects on the observed infrared spectra and that the computations are performed within the harmonic approximation. The spectrum of the KBr pellet can be now compared with the RAIRS spectrum of a 50 nm PTCDA film deposited on a smooth reflecting silver mirror. The RAIRS results are shown in Figure 7.9, where the calculated vibrational intensities for the b_{3u} species are also included to facilitate the assignment. According to the surface selection rules, only the vibrational modes having nonzero dipole moment derivative components perpendicular to the surface should be active in RAIRS. Hence, from the RAIRS spectrum of the PTCDA molecule, it can be extracted that it is preferentially oriented with its x-axis (out-of-molecular plane) perpendicular to the metal surface, a flat-on molecular orientation. Similar results were obtained for PTCDA films of 20 nm mass thickness [104]. The fact that the in-plane vibrations are still seen is an indication that in this relatively thick film (50 nm mass thickness) there is a certain degree of randomness.

The SEIRA spectrum was also obtained by depositing a 50 nm PTCDA film on a silver island film (15 nm mass thickness of Ag), for which an AFM image and the corresponding plasmon absorption are shown in Figure 7.10. The experimental conditions for the deposition of the

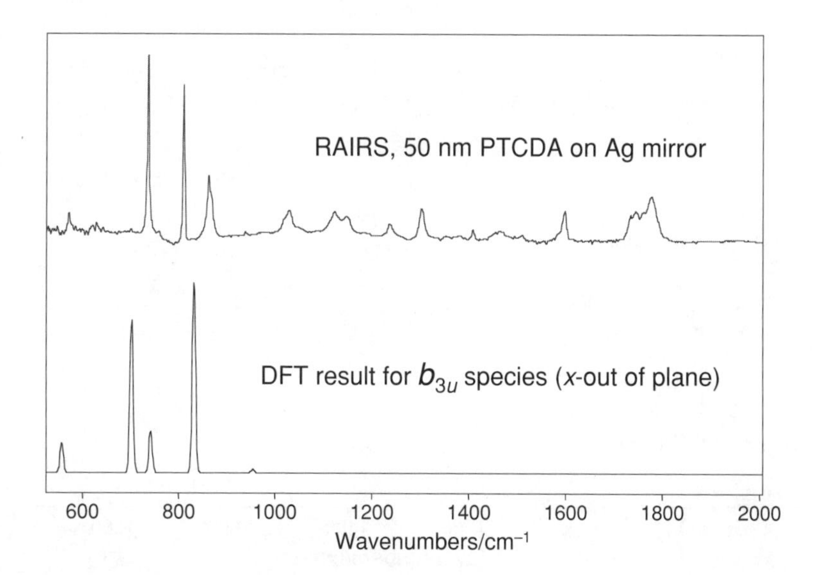

Figure 7.9 Reflection–absorption infrared spectrum of a 50 nm mass thickness evaporated film of PTCDA deposited on a smooth reflecting silver film. The bottom trace corresponds to the calculated vibrational infrared intensities for the b_{3u} species of symmetry. Reproduced from R.F. Aroca, D.J. Ross and C. Domingo, Surface-enhanced infrared spectroscopy, *Appl. Spectrosc.* 2004, **58**, 324A–338A by permission of Applied Spectroscopy

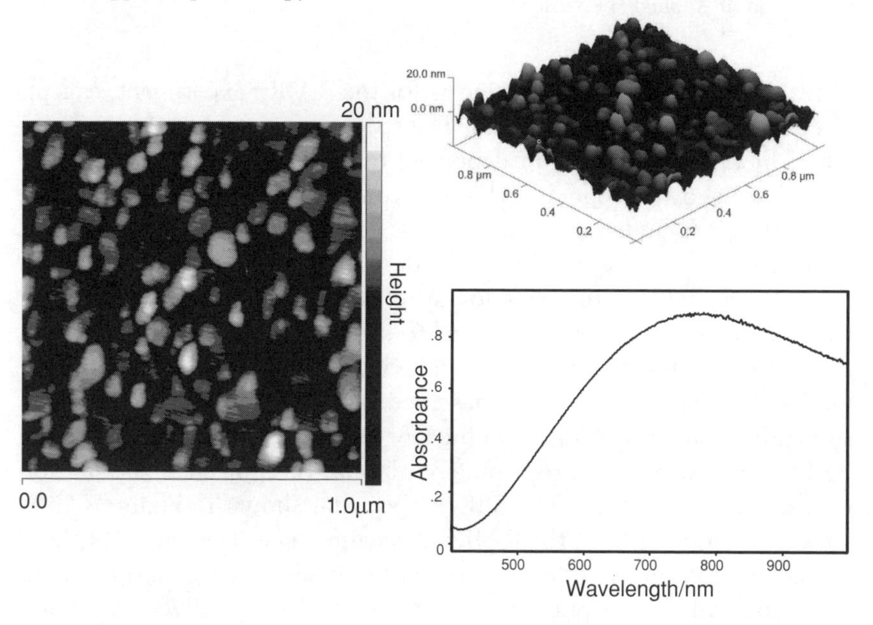

Figure 7.10 Atomic force microscopy image and UV–Visible absorption spectrum of a 15 nm mass thickness evaporated Ag film. Reproduced from R.F. Aroca, D.J. Ross and C. Domingo, Surface-enhanced infrared spectroscopy, *Appl. Spectrosc.* 2004, **58**, 324A–338A by permission of Applied Spectroscopy

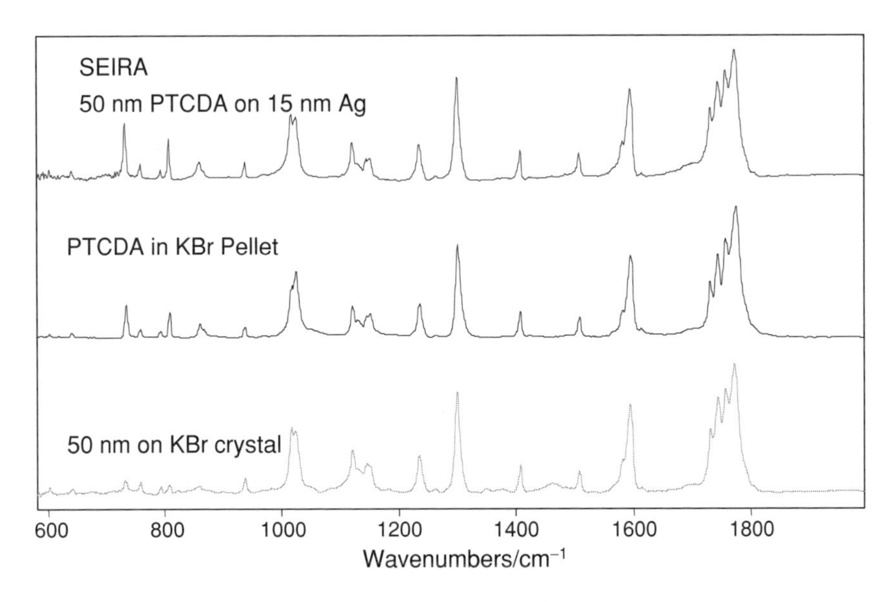

Figure 7.11 Transmission surface-enhanced infrared spectrum of a 50 nm PTCDA film on silver islands (top), transmission FTIR spectrum of PTCDA in a KBr pellet (middle) and transmission FTIR spectrum of a 50 nm evaporated PTCDA film on a KBr crystal (bottom) Reproduced from R.F. Aroca, D.J. Ross and C. Domingo, Surface-enhanced infrared spectroscopy, *Appl. Spectrosc.* 2004, **58**, 324A–338A by permission of Applied Spectroscopy

SEIRA metal surface are the same as for the RAIRS experiment, except, of course, for the mass thickness. The spectroscopic results for a 50 nm mass thickness PTCDA film deposited on silver islands and on a KBr crystal are shown in Figure 7.11. For comparison, the spectrum of the KBr pellet is included as a reference. If one assumes the same molecular orientation in the PTCDA films deposited on a smooth silver film and KBr crystal, the transmission spectrum of the film on KBr, where the electric field is polarized parallel to the surface, should give a strong absorption for the in-plane mode and a weak signal for the out-of-plane modes (b_{3u}). Since the latter is observed (Figure 7.11), the transmission spectrum of the PTCDA film on the KBr crystal is in agreement with the RAIRS results, pointing to a preferential flat-on molecular orientation in the evaporated films. The SEIRA spectrum shown in Figure 7.11 is strikingly different from the RAIRS spectrum and clearly does not show the same selection rules. Since in the infrared spectral region one would also expect the local optical field on the surface of the silver island film to be perpendicular to the surface (as in RAIRS), the simplest explanation would be to assume that the PTCDA molecules are not oriented flat-on the metal island, and instead there is a random distribution of orientations and the vibrational intensities are much closer to that of the free molecule (or KBr spectrum).

Since the films used in the previous experiments were relatively 'thick', the experiments were repeated using a PTCDA film of 10 nm mass thickness. The transmission spectrum of the neat PTCDA film on the KBr crystal and the SEIRA spectrum of the 10 nm PTCDA film on 15 nm silver island film were recorded under identical experimental conditions using a Bomem DA3 vacuum bench instrument. The RAIRS spectrum of the 10 nm PTCDA film on smooth silver mirror was recorded with a Bruker instrument with p-polarized light. The results are shown in Figure 7.12. It can be seen that the selection rules observed in the RAIRS spectrum unmistakably point to a flat-on molecular orientation of PTCDA with a strong relative intensity for the out-of-plane b_{3u}. Notably, the SEIRA spectrum does not follow the RAIRS pattern. However, the relative intensity of the out-of-plane modes has increased with decreasing film thickness, hinting at a proportional increase of PTCDA molecules oriented flat-on the silver islands.

The PTCDA case presented here is of interest for researchers trying to extract molecular orientation information using SEIRA. In such cases,

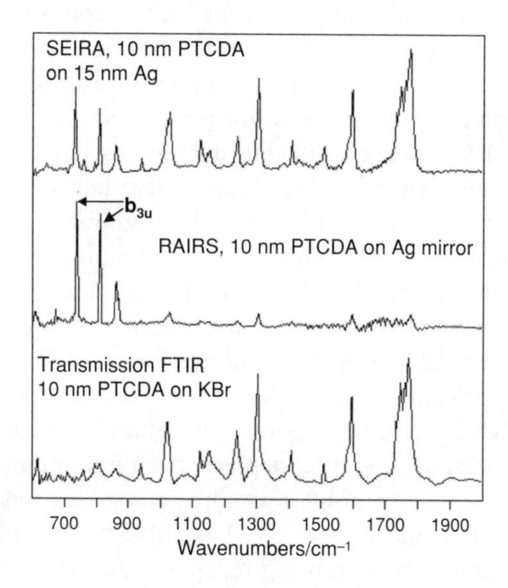

Figure 7.12 Transmission SEIRA spectrum of the 10 nm PTCDA film evaporated on a 15 nm silver island film (top), RAIRS spectrum of the 10 nm PTCDA film evaporated on a smooth silver mirror (middle), and transmission FTIR spectrum of the 10 nm PTCDA evaporated film on the surface of a KBr crystal (bottom) Reproduced from R.F. Aroca, D.J. Ross and C. Domingo, Surface-enhanced infrared spectroscopy, *Appl. Spectrosc.* 2004, 58, 324A–338A by permission of Applied Spectroscopy

it is advisable to carry out the RAIRS experiments [107,108] with the same species as a reference point for the SEIRA results. In conclusion, the observation of strict adherence to surface selection rules in RAIRS may not be mimicked by the SEIRA spectrum.

7.5 APPLICATIONS OF SEIRA

7.5.1 SEIRA of Ultrathin Films

SEIRA, like SERS, is a powerful technique for structural characterization of ultrathin films and well-ordered monolayers on metal surfaces. Thin films at interfaces have been prepared with different procedures and developed for various applications. The fabrication and characterization of ultrathin films is an exciting area of research [64,109,110] where some of the most interesting subjects are (1) bilayers and monolayers at the liquid–liquid interface, (2) adsorption monolayers and Langmuir (water-insoluble) monolayers at the air–water interface, (3) adsorption films and SAMs at the liquid–solid interface and (4) LB films, cast (deposit) films and spin-coat films at the air–solid interface. The molecular organization in these thin films depends on the conditions of preparation and can be extracted using vibrational techniques. Among the many SEIRA applications to films and interfaces one finds studies about molecular organization of monolayers of porphyrin derivatives [111,112] and of azamacrocycles and their metallic derivatives [113,114] at the air–solid interface. Dendrimers architectures adsorbed on SAM modified gold surfaces, have also been investigated using SEIRA [115,116]. Similarly, bifunctional molecules that have interesting technological applications are also among the SEIRA targets. For instance, adlayers of 2,2′-bypiridine have been formed with electrochemical techniques on Cu and investigated with SEIRA at solid–liquid interfaces [117]. These molecules could work as molecular electric devices, as their reversible orientation can be controlled by electrode potentials. The electrochemical adsorption of 4,4′-Bypiridine (4,4′-BP), its phase behavior and its coadsorption with interfacial water on gold (thin-film) electrodes has been studied using *in situ* SEIRA and *ex situ* STM [118]. The 4,4′-BP bifunctional nonchelating ligand acts as a bridging spacer and coordination unit in 3D and/or 2D supramolecular lattices with novel electric and magnetic properties.

Recently, characterization using surface-enhanced vibrational spectroscopy, SEIRA and SERS, has begun to emerge as demonstrated in the study of nanoarrayed superstructures formed by adsorption of 1,4-phenylenediisocyanide (1,4-PDI) on gold nanoparticles [119]. 1,4-PDI is adsorbed on gold via the carbon lone-pair electrons of one isocyanide group assuming a vertical orientation with respect to the gold substrate. The pendent isocyanide group is further identified by SEIRA, AFM and QCM to react with Au nanoparticles. SEIRA spectroscopy also revealed that 1,4-PDI molecules could be newly adsorbed on those Au nanoparticles, implying that nanoarrayed electrodes could be fabricated using 1,4-PDI as the conducting wires of Au nanoparticles.

7.5.2 Surface Photochemistry and Catalytic Reactions

Given the catalytic activity of some metal particles, SEIRA is a new analytical tool in catalysis. SEIRA has been successfully used to help explain the photoenhanced hydrophilicity of the photocatalyst TiO_2 [120], and later SEIRA was applied to the *in situ* observation of surface products during the photooxidation of gas-phase *n*-decane on TiO_2 films coated with Au [121] and Pt [122]. Similarly, the photocatalytic decomposition of acetic acid, in both the liquid and vapor phase, using TiO_2 has been investigated with SEIRA [123]. The feasibility of *in situ* ATR–SEIRA spectroscopy for the study of liquid-phase heterogeneous catalysis by platinum metals has also been reported [124].

7.5.3 Electrochemistry

SEIRA, most commonly performed using the ATR Kretschmann configuration, is a successful *in situ* surface-sensitive technique for electrochemical dynamic studies, and several reviews on such applications have been published [64,125]. A discussion of the practical advantages of the ATR Kretschmann configuration in electrochemistry has been published [64]. In summary, free mass transport, less interference from the solution and higher sensitivity due to SEIRA enhancement give SEIRA a unique high sensitivity for infrared measurements on time-scales of fast cyclic voltammetry. Indeed, the time resolution for measurements of CO adsorbed on Pt is in the range of microseconds with the use of a step-scan interferometer, and a sensitivity 30–75 times higher than with conventional IR spectroscopy, including the effect of enhancement due to the rough electrode, which is of the order of seven [126]. Furthermore, thanks to developments in FTIR instrumentation and data analysis techniques such as 2D-IR, SEIRA has allowed the study of electrode dynamics that is not readily accessible by conventional electrochemical techniques [127]. Recently, thanks to the advantages of SEIRA, formate has been detected, for the first time, as an active intermediate species in the non-CO pathway of the electro-oxidation of methanol on platinum electrodes [128]. Although the catalytic electro-oxidation of methanol (and methane) on Pt or Pt-based metal electrodes had been extensively studied owing to their potential use in low-temperature fuel cells, the detection of formate strongly suggests that many previous reaction mechanisms may have to be reconsidered. SEIRA experiments have also help to show that formate is the active intermediate in the electrooxidation of formaldehyde on Pt electrodes [129]. *In situ* SEIRA studies on the anodic oxidation of methane at Pt, Au, Pd, Ru and Rh electrodes [19] have revealed that it is possible to activate methane at room temperature using noble metals as electrocatalysts, and that Pt and Ru seems to be the most active ones for such purpose.

7.5.4 Analytical Applications

SEIRA is a surface-sensitive spectroscopic technique that has the great advantage of profiting from a vast body of the vibrational data, collected, understood and classified in libraries for all three state of matter. In addition, a modest enhancement factor (10–100) makes IR spectroscopy an attractive and powerful analytical tool. The cross-sections of infrared absorption are orders of magnitude larger than Raman cross-sections. For instance, the cross-section σ for absorption in the infrared region is ca $10^{-21}\,\text{cm}^2$. The absolute Raman cross-section for the $666\,\text{cm}^{-1}$ mode of chloroform has been determined [130], using the 532.0 nm laser line, as $\sigma_R = (0.660 \pm 0.1) \times 10^{-28}\,\text{cm}^2$. Therefore, the sensitivities of 'average' SERS and SEIRA are comparable and both are currently under development as promising optical sensor technologies.

The analytical applications of SEIRA as a surface-sensitive spectroscopic technique require a thorough examination of metal–molecule interactions and polarization effects that may give rise to a distinct vibrational spectrum, in any case, very different from that of the parent molecule. Peak positions and relative intensities in the enhanced spectra may be different from those of normal spectra of the same molecules, and the deviation is larger for strongly chemisorbed adsorbates. The latter means that the databases of normal IR spectra cannot be used directly for automatic identification of compounds through their SEIRA spectra and that new databases must be built. The enhancement is short range, being most effective for molecules on or near the metal surface. The relative intensity in SEIRA is not a linear function of the amount of molecules.

SEIRA in trace analysis has been used as a detector for flow-injection system and applied in the analysis of environmentally hazardous chemicals in waste water [65]. Recently, good results have been reported by coupling SEIRA to LC [131] and to GC, employing silver islands on ZnSe as SEIRA-active surface [132]. Detection limits 10 times lower in comparison with GC–FTIR have been reported for chemisorbed molecules. As an immediate objective, the equipment for GC– and HPLC–SEIRA recently developed in Griffiths's group will be tested for the analysis of drugs in hair, as an alternative to urinalysis.

Thiram and ziram, two dithiocarbamate fungicides of potential aggressive environmental impact, have been the subject of concurrent SEIRA and SERS studies, using the same 'enhancing substrate' [133]. For these compounds, SEIRA is the preferred enhanced technique. Several biological applications have also been reported using SEIRA, and it has been tested for various bioanalytical purposes including immmunoassays [68,134]. Biosensors use antibodies or enzymes, immobilized on a platform, to interact selectively with antigens or substrates, followed by one of several possible transduction mechanisms to detect that interaction. The reported SEIRA-based biosensors have used colloidal gold to immobilize antibodies, either for *Salmonella* [68] or for staphylococal protein A [134], employing external reflection and ATR techniques, respectively. In a different application, SEIRA microspectroscopy was used for localization of bacteria on geological material surfaces and their ulterior investigation [135]. In that case, gold was evaporated on the samples.

Relevant information about the adsorption and orientation of nucleic acid bases on metal surfaces has been obtained from SEIRA studies of thymine on silver island films [136], cytosine [137] and uracil [138] on gold electrodes. On the other hand, SEIRA studies of the structure of nucleic acids and phospholipids from tumor cells, both sensitive- and drug-resistant [139–141], seem to have generated new expectations about the possibility of using SEIRA for diagnostic criteria in cancer research. The reason is that this method enhances a set of the bands that are impossible to observe in conventional infrared spectroscopy, but that are crucial for determining nucleic acid structural peculiarities from tumor tissues and nucleic acid interactions with anticancer drugs. SEIRA data on DNA interactions with single-walled carbon nanotubes (SWCNTs) [142] could be explained by the model of wrapping of nucleic acid molecules around carbon nanotubes proposed previously. This is an interesting result because a similar situation seems to occur in chromosomes during DNA assembling by histones.

Adsorption of protein-rich vesicles on Ag-cluster coated Ge crystals in an aqueous environment has been investigated with SEIRA [143]. Nicotinic acetylcholine receptor, the structurally best-characterized prototype of the ligand-gate neuroreceptor, was the membrane protein contained in the vesicles. The SEIRA technique has thus allowed the detection of small and specific changes in the structure of biomembranes, changes practically impossible to detect by normal FTIR spectroscopy.

The electrochemically induced oxidation and reduction process of a monolayer of horse heart cytochrome *c*, protein that mediates single-electron transfer between the integral membrane protein complexes of the respiratory chain, has been followed by SEIRA [144]. In this case the SEIRA-active substrate was rough gold modified with a SAM of mercaptopropionic acid (MPA). The results obtained demonstrate that minute enzymatic changes of a protein can be studied at the level of a monolayer using SEIRA. SEIRA has been applied in the analysis of plasma-modified polymer surfaces [145] and in the study of the adhesion mechanisms of functional monomers used for the surface treatment of dental alloys, on the basis of their molecular structural information [146]. Other interesting materials, fullerenes, have been also explored with SEIRA [147].

SEIRA is the complement of SERS forming the experimental part of surface-enhanced vibrational spectroscopy. The enhancement factors in the absorption spectra are very modest in comparison with those observed in inelastic scattering. However, infrared absorption cross-sections are orders of magnitude higher than the corresponding Raman cross-sections, and a modest increase could have an enormous effect in practical applications. The vast body of infrared data guarantees a broad range of analytical applications and in particular development as a surface analytical technique.

REFERENCES

[1] M. Kerker (ed.), *Selected Papers on Surface-enhanced Raman Scattering*. SPIE, Bellingham, WA, 1990.

[2] M. Moskovits, *Rev. Mod. Phys.* 1985, **57**, 783–826.

[3] H. Chew, D.S. Wang and M. Kerker, Surface enhancement of coherent anti-Stokes Raman scattering by colloidal spheres, *J. Op. Soc. Am. B* 1984, **1**, 56–66.

[4] K. Kneipp, H. Kneipp, I. Itzkan, R.R. Dasari and M.S. Feld, Surface-enhanced non-linear Raman scattering at the single-molecule level, *Chem. Phys.* 1999, **247**, 155–162.

[5] G.C. Schatz and R.P. Van Duyne, in J.M. Chalmers and P.R. Griffiths (eds), *Handbook of Vibrational Spectroscopy*, John Wiley & Sons, Ltd, Chichester, 2002, pp. 759–774.

[6] A. Hartstein, J.R. Kirtley and J.C. Tsang, Enhancement of the infrared absorption from molecular monolayers with thin metal overlayers, *Phys. Rev. Lett.* 1980, **45**, 201–204.

[7] T.R. Jensen, R.P. Van Duyne, S.A. Johnson and V.A. Maroni, Surface-enhanced infrared spectroscopy: a comparison of metal island films with discrete and nondiscrete surface plasmons, *Appl. Spectrosc.* 2000, **54**, 371–377.

[8] C.F. Eagen, Nature of the enhanced optical absorption of dye-coated Ag islands, *Appl. Opt.* 1981, **20**, 3035–3042.

[9] H. Muraki, T. Ito and M. Hiramatsu, Enhanced absorption of Langmuir–Blodgett monolayers on silver island films, *Thin Solid Films* 1989, **179**, 509–513.

[10] R. Hillenbrand, T. Taubner and F. Keilmann, Phonon-enhanced light-matter interactions at the nanometre scale, *Nature* 2002, **418**, 159–162.

[11] P.C. Andersen and K.L. Rowlen, Brilliant optical properties of nanometric noble metals spheres, rods, and apertures arrays, *Appl. Spectrosc* 2002, **56**, 124A–135A.

[12] M.S. Anderson, Enhanced infrared absorption with dielectric nanoparticles, *Appl. Phys. Lett.* 2003, **83**, 2964–2966.

[13] M. Osawa, Dynamic processes in electrochemical reactions studied by surface-enhanced infrared absorption spectroscopy (SEIRAS), *Bull. Chem. Soc. Jpn.* 1997, **70**, 2861–2880.

[14] Y. Nishikawa, K. Fujiwara, K. Ataka and M. Osawa, Surface-enhanced infrared external reflection spectroscopy at low reflective surfaces and its application to surface analysis of semiconductors, glasses, and polymers, *Anal. Chem.* 1993, **65**, 556–562.

[15] E. Johnson and R. Aroca, Surface-enhanced infrared spectroscopy of monolayers, *J. Phys. Chem.* 1995, **99**, 9325–9330.

[16] R. Kellner, B. Mizaikoff, M. Jakusch, H.D. Wanzenbock and N. Weissenbacher, Surface-enhanced vibrational spectroscopy: a new tool in chemical IR sensing? *Appl. Spectrosc.* 1997, **51**, 495–503.

[17] A.E. Bjerke, P.R. Griffiths and W. Theiss, Surface-enhanced infrared absorption of CO on platinized platinum, *Anal. Chem.* 1999, **71**, 1967–1974.

[18] R. Aroca and B. Price, A new surface for surface-enhanced infrared spectroscopy: tin island films, *J. Phys. Chem. B* 1997, **101**, 6537–6541.

[19] F. Hahn and C.A. Melendres, Anodic oxidation of methane at noble metal electrodes: an '*in situ*' surface enhanced infrared spectroelectrochemical study, *Electrochim. Acta* 2001, **46**, 3525–3534.

[20] M.S. Zheng, S.G. Sun and S.P. Chen, Abnormal infrared effects and electrocatalytic properties of nanometer scale thin film of Pt–Ru alloys for CO adsorption and oxidation, *J. Appl. Electrochem.* 2001, **31**, 749–757.

[21] A. Otto, I. Mrozek, H. Grabhorn and W. Akemann, Surface-enhanced Raman scattering, *J. Phys. Condens. Matter* 1992, **4**, 1143–1212.

[22] A. Campion and P. Kambhampati, Surface-enhanced Raman scattering, *Chem. Soc. Rev.* 1998, **27**, 241–250.

[23] R.F. Aroca, D.J. Ross and C. Domingo, Surface-enhanced infrared spectroscopy, *Appl. Spectrosc.* 2004, **58**, 324A–338A.

[24] J. Sukmanowski, J.-R. Viguié, B. Nölting and F.X. Royer, Light absorption enhancement by nanoparticles, *J. Appl. Phys.* 2005, **97**, 104332–104337.

[25] M. Osawa and M. Ikeda, Surface-enhanced infrared absorption of *p*-nitrobenzoic acid deposited on silver island films: contributions of electromagnetic and chemical mechanisms, *J. Phys. Chem.* 1991, **95**, 9914–9919.

[26] D. Ross and R. Aroca, Effective medium theories in surface enhanced infrared spectroscopy: the pentacene example, *J. Chem. Phys.* 2002, **117**, 8095–8103.

[27] E.A. Coronado and G.C. Schatz, Surface plasmon broadening for arbitrary shape nanoparticles: a geometrical probability approach. *J. Chem. Phys.* 2003, **19**, 3926–3934.

[28] D.K. Lambert, Electric field induced change of adsorbate vibrational line strength, *J. Chem. Phys.* 1991, **94**, 6237–6242.

[29] O. Krauth, G. Fahsold and A. Lehmann, Surface-enhanced infrared absorption? *Surf. Sci.* 1999, **433–435**, 79–82.

[30] O. Krauth, G. Fahsold, N. Magg and A. Pucci, Anomalous infrared transmission of adsorbates on ultrathin metal films: Fano effect near the percolation threshold, *J. Chem. Phys.* 2000, **113**, 6330–6333.

[31] M. Sinther, A. Pucci, A. Otto, A. Priebe, S. Diez and G. Fahsold, enhanced infrared absorption of SERS active lines of ethylene on Cu, *Phys. Status Solidi A* 2001, **188**, 1471–1476.

[32] D.C. Langreth, Energy transfer at surfaces: asymmetric line shapes and the electron–hole pair mechanism, *Phys. Rev. Lett.* 1985, **54**, 126–129.

[33] A. Priebe, G. Fahsold and A. Pucci, Surface enhanced infrared absorption of CO on smooth iron ultrathin films, *Surf. Sci.* 2001, **482–485**, 90–95.

[34] A. Priebe, M. Sinther, G. Fahsold and A. Pucci, The correlation between film thickness and adsorbate line shape in surface enhanced infrared absorption, *J. Chem. Phys.* 2003, **119**, 4887–4890.

[35] K. Kneipp, Y. Wang, H. Kneipp, L.T. Perelman, I. Itzkan, R. Dasari and M.S. Feld, *Phys. Rev. Lett.* 1997, **78**, 1667–1670.

[36] S. Nie and S.R. Emory, Probing single molecules and single nanoparticles by surface–enhanced Raman scattering, *Science* 1997, **275**, 1102–1106.

[37] G.C. Papavassiliou, Optical properties of small inorganic and organic metal particles, *Prog. Solid State Chem.* 1979, **12**, 185–271.

[38] D.L. Feldheim and C.A. Foss (eds), *Metal Nanoparticles. Synthesis, Characterization and Applications*, Marcel Dekker, New York, 2002.

[39] E.A. Coronado and G.C. Schatz, Surface plasmon broadening for arbitrary shape nanoparticles: a geometrical probability approach, *J. Chem. Phys.* 2003, **119**, 3926–3934.

[40] M. Osawa, Surface-enhanced infrared absorption, *Top. Appl. Phys.* 2001, **81**, 163–187.

[41] P.W. Barber, R.K. Chang and H. Massoudi, Electrodynamic calculations of the surface-enhanced electric intensities on large Ag spheroids, *Phys. Rev. B* 1983, **27**, 7251.

[42] G.C. Schatz and R.P. Van Duyne, in J.M. Chalmers and P.R. Griffiths (eds), *Handbook of Vibrational Spectroscopy*, John Wiley & Sons, Inc., New York, 2002, p. 759–774.

[43] E.J. Zeman and G.C. Schatz, An accurate electromagnetic theory study of surface enhancement factors of Ag, Au, Cu, Li, Na, Ga, In, Zn, Cd, *J. Phys. Chem.* 1987, **91**, 634–643.

[44] J.C. Maxwell-Garnet, Colours in Metal Glasses and in Metallic Films. *Trans. R. Soc. London* 1904, **203**, 385–420.

[45] M. Osawa, K.-I. Ataka, K. Yoshii and Y. Nishikawa, Surface-enhanced infrared spectroscopy: the origin of the absorption enhancement and band selection rule in the infrared spectra of molecules adsorbed on fine metal particles. *Appl. Spectrosc.* 1993, **47**, 1997–1502.

[46] C.F Bohren and D.R. Huffman, *Absorption and Scattering of Light by Small Particles*, John Wiley & Sons, Inc., New York, 1983.

[47] H. Fujiwara, J. Koh, P.I. Rovira and R.W. Collins, Assessment of effective-medium theories in the analysis of nucleation and microscopic surface roughness evolution for semiconductor thin films. *Phys. Rev. B* 2000, **61**, 10832–10844.

[48] D.J. Bergman and D. Stroud, Physical properties of macroscopically inhomogeneous media, *Solid State Phys.* 1992, **46**, 148–270.

[49] D.J. Bergman, Nonlinear behaviour and 1/f noise near a conductivity threshold: effects of local microgeometry, *Phys. Rev. B* 1989, **39**, 4598–4609.

[50] D.J. Bergman and H.J. Dunn, Bulk effective dielectric constant of a composite with a periodic microgeometry, *Phys. Rev. B* 1992, **45**, 13262–13271.

[51] E.C. Stoner. The demagnetizing factors for ellipsoids, *Philos. Mag.* 1945, **36**, 803–821.

[52] J.A. Osborn, Demagnetizing factors of the general ellipsoid, *Phys. Rev. B* 1945, **67**, 351–357.

[53] D.J. Jackson, *Classical Electrodynamics*, Jonh Wiley & Sons, Inc., New York, 1975.

[54] S. Normann, T. Andersson, C.G. Granqvist and O. Hunderi, Optical properties of discontinuous gold films, *Phys. Rev. B* 1978, **18**, 674–695.

[55] E.D. Palik (ed.), *Handbook of Optical Constants of Solids*, Academic Press, New York, 1985.

[56] A. Roseler and E.H. Korte, The optical constant of metallic island films as used for surface enhanced infrared absoprtion, *Thin Solid Films* 1998, **313–314**, 732–736.

[57] S. Yoshida, T. Yamaguchi and A. Kimbara, Optical properties of aggregated silver films, *J. Opt. Soc. Am.* 1971, **61**, 62.

[58] R.R. Bilboul, A note on the permittivity of a double-layer ellipsoid, *Br. J. Appl. Phys. (J. Phys. D)* 1969, **2**, 921–923.

[59] E.J. Zeman and G.C. Schatz, An accurate electromagnetic theory study of surface enhancement factors for silver, gold, copper, lithium, sodium, aluminum, gallium, indium, zinc, and cadmium, *J. Phys. Chem.* 1987, **91**, 634–643.

[60] S. Heutz, G. Salvan, S.D. Silaghi, T.S. Jones and D.R.T. Zahn, Raman scattering as a probe of crystallinity in PTCDA and H2Pc single-layer and double-layer thin film heterostructures, *J. Phys. Chem. B* 2003, **107**, 3782–3788.

[61] S.B. Chaney, S. Shanmukh, R.A. Dluhy and Y.-P. Zhao, *Appl. Phys. Lett.* 2005, **87**, 31908.

[62] R.A. Dluhy and C. Leverette, personal communication, 2005.

[63] Z. Zhang and T. Imae, Study of surface-enhanced infrared spectroscopy: 2. Large enhancement achieved through metal–molecule–metal sandwich configurations, *J. Colloid Interface Sci.* 2001, **233**, 107–111.

[64] M. Osawa, in J.M. Chalmers and P.R. Griffiths (eds), *Handbook of Vibrational Spectroscopy*, John Wiley & Sons, Inc., New York, 2002, pp. 785–799.

[65] H.D. Wanzenboeck, B. Mizaikoff, N. Weissenbacher and R. Kellner, Surface enhanced infrared absorption spectroscopy (SEIRA) using external reflection on low-cost substrates, *Fresenius' J. Anal. Chem.* 1998, **362**, 15–20.

[66] N. Al-Rawashdeh and C.A. Foss Jr, UV/visible and infrared spectra of polyethylene/nanoscopic gold rod composite films: effects of gold particle size, shape and orientation, *Nanostruct. Mater.* 1997, **9**, 383–386.

[67] S.Y. Kang, I.C. Jeon and K. Kim, *Appl. Spectrosc.* 1998, **52**, 278.

[68] A.A. Kamnev, L.A. Dykman, P.A. Tarantilis and M.G. Polissiou, Spectroimmunochemistry using colloidal gold bioconjugates, *Biosci. Rep.* 2002, **22**, 541–547.

[69] C. Domingo, J.V. García-Ramos, S. Sánchez-Cortes and J.A. Aznárez, SERS and SEIR joint studies on gold, silver and copper nanostructures, in *Proceedings of XVIII ICORS*, John Wiley & Sons, Inc., New York, 2002, pp. 295–296.

[70] K.C. Grabar, R.G. Freeman, M.B. Hommer and M.J. Natan, Preparation and characterization of Au colloid monolayers, *Anal. Chem.* 1995, **67**, 735–745.

[71] E. Garcia-Caurel, E. Bertran and A. Canillas, Application of FTIR phase-modulated ellipsometry to the characterization of thin films on surface-enhanced IR absorption active substrates, *Thin Solid Films*, 2001, **398–399**, 99–103.

[72] Y. Zhu, H. Uchida and M. Watanabe, Oxidation of carbon monoxide at a platinum film electrode studied by Fourier transform infrared spectroscopy with attenuated total reflection technique, *Langmuir*, 1999, **15**, 8757–8764.

[73] H. Miyake, S. Ye and M. Osawa, Electroless deposition of gold thin films on silicon for surface-enhanced infrared spectroelectrochemistry, *Electrochem. Commun.* 2002, **4**, 973–977.

[74] J. Yang and S.-H. Chen, Development of electrode-less plating method for silver film preparations for surface-enhanced infrared absorption measurements, *Appl. Spectrosc.* 2001, **55**, 399–406.

[75] A. Rodes, J.M. Orts, J.M. Perez, J.M. Feliu and A. Aldaz, *Electrochem. Commun.* 2003, **5**, 56–60.

[76] A. Miki, S. Ye and M. Osawa, Surface-enhanced IR absorption on platinum nanoparticles: an application to real-time monitoring of electrocatalytic reactions, *Chem. Commun.* 2002, 1500–1501.

[77] H. Miyake, S. Ye and M. Osawa, 204th Meeting, The Electrochemical Society, Orlando, FL, 2003, Abstract 15.

[78] H. Shiroishi, Y. Ayato, K. Kunimatsu and T. Okada, Study of adsorbed water on Pt during methanol oxidation by ATR–SEIRAS (surface-enhanced infrared absorption spectroscopy), *J. Electroanal. Chem.* 2005, **581**, 132–138.

[79] K. Kunimatsu, T. Senzaki, M. Tsushima and M. Osawa, A combined surface-enhanced infrared and electrochemical kinetics study of hydrogen adsorption and evolution on a Pt electrode, *Chem. Phys. Lett.* 2005, **401**, 451–454.

[80] N. Goutev and M. Futamata, Attenuated total reflection surface-enhanced infrared absorption spectroscopy of carboxyl terminated self-assembled monolayers on gold, *Appl. Spectrosc.* 2003, **57**, 506–513.

[81] J.A. Seelenbinder and C.W. Brown, Comparison of organic self-assembled monolayers as modified substrates for surface-enhanced infrared absorption spectroscopy, *Appl. Spectrosc.* 2002, **56**, 295–299.

[82] J.A. Seelenbinder, C.W. Brown and D.W. Urish, Self-assembled monolayers of thiophenol on gold as a novel substrate for surface-enhanced infrared absorption, *Appl. Spectrosc.* 2000, **54**, 366–370.

[83] C. Domingo, J.V. García-Ramos, S. Sánchez-Cortés and J.A. Aznarez, *J. Mol. Struct.* 2003, **661–662**, 419.
[84] G.T. Merklin and P.R. Griffiths, *Langmuir* 1997, **13**, 6159–6161.
[85] T. Yoshidome and S. Kamata, Surface enhanced infrared spectra with the use of the Pb film and its application to the microanalyses, *Anal. Sci.* 1997, **13**, 351–354.
[86] Y. Nakao and H. Yamada, Enhanced infrared ATR spectra of surface layers using metal films, *Surf. Sci.* 1986, **176**, 578–592.
[87] G.-Q. Lu, S.-G. Sun, L.-R. Cai, S.-P. Chen, Z.-W. Tian and K.-K. Shiu, *In situ* FTIR spectroscopic studies of adsorption of CO, SCN−, and poly(*o*-phenylenediamine) on electrodes of nanometer thin films of Pt, Pd, and Rh: abnormal infrared effects (AIREs), *Langmuir* 2000, **16**, 778–786.
[88] S. Zou, R. Gomez and M.J. Weaver, Nitric oxide and carbon monoxide adsorption on polycrystalline iridium electrodes: a combined Raman and infrared spectroscopic study, *Langmuir* 1997, **13**, 6713–6721.
[89] M. Watanabe, Y. Zhu and H. Uchida, Oxidation of CO on a Pt–Fe alloy electrode studied by surface enhanced infrared reflection–absorption spectroscopy, *J. Phys. Chem. B* 2000, **104**, 1762–1768.
[90] A.E. Bjerke and P.R. Griffiths, Surface-enhanced infrared absorption spectroscopy of *p*-nitrothiophenol on vapor-deposited platinum films, *Appl. Spectrosc.* 2002, **56**, 1275–1280.
[91] S.-G. Sun, in C. Vayenas (ed), *Catalysis and Electrocatalysis at Nanoparticle Surfaces*, Marcel Dekker, New York, 2003, pp. 785–826.
[92] E.M. Kosower G. Markovich, Y. Raichlin, G. Borz and A. Katzir, Surface-enhanced infrared absorption and amplified spectra on planar silver halide fiber, *J. Phys. Chem. B*, 2004, **108**, 12633–12636.
[93] E.M. Kosower, G. Markovich and G. Borz, Surface-enhanced infrared absorption of *p*-nitrobenzoic acid on planar silver halide fiber, *J. Phys. Chem. B* 2004, **108**, 12873–12876.
[94] A.S. Davydov, *Quantum Mechanics*, Pergamon Press, Oxford, 1965.
[95] J.C. Decius and R.M. Hexter, *Molecular Vibrations in Crystals*, McGraw-Hill, New York, 1977.
[96] G. Herzberg, *Molecular Spectra and Molecular Structure. II. Infrared and Raman Spectra of Polyatomic Molecules*, Van Nostrand, Princeton, NJ, 1945.
[97] W.B. Person and G. Zerbi (eds), *Vibrational Intensities in Infrared and Raman Spectroscopy*, Elsevier, Amsterdam, 1982.
[98] G. Herzberg, *Molecular Spectra and Molecular Structure. III. Electronic Spectra and Electronic Structure of Polyatomic Molecules*, Van Nostrand, New York, 1966.
[99] R.G. Greenler, *J. Chem. Phys.* 1966, **44**, 310–315.
[100] H.A. Pearce and N. Sheppard, *Surf. Sci.* 1976, **59**, 205.
[101] R.M. Hexter and M.G. Albrecht, *Spectrochim. Acta Part A* 1979, **35**, 233–251.
[102] M.C. Gerstenberg, F. Schreiber, T.Y.B. Leung, G. Bracco, S.R. Forrest and G. Scoles, Organic semiconducting thin film growth on an organic substrate: 3,4,9,10-perylenetetracarboxylic dianhydride on a monolayer of decanethiol self-assembled on Au(111), *Phys. Rev. B* 2000, **61**, 7686–7691.
[103] K. Akers, R. Aroca, A.M. Hor and R.O. Loutfy, Molecular organization in perylene tetracarboxylic dianhydride films, *J. Phys. Chem.* 1987, **91**, 2954–2959.
[104] R. Aroca and S. Rodriguez-Llorente, Surface-enhanced vibrational spectroscopy, *J. Mol. Struct.* 1997, **408/409**, 17–22.

[105] M.J. Frisch, G.W. Trucks, H.B. Schlegel, G.E. Scuseria, M.A. Robb, J.R. Cheeseman, V.G. Zakrzewski, J.A. Montgomery Jr, R.E. Stratmann, J.C. Burant, S. Dapprich, J.M. Millam, A.D. Daniels, K.N. Kudin, M.C. Strain, O. Farkas, J. Tomasi, V. Barone, M. Cossi, R. Cammi, B. Mennucci, C. Pomelli, C. Adamo, S. Clifford, J. Ochterski, G.A. Petersson, P.Y. Ayala, Q. Cui, K. Morokuma, D.K. Malick, A.D. Rabuck, K. Raghavachari, J.B. Foresman, J. Cioslowski, J.V. Ortiz, B.B. Stefanov, G. Liu, A. Liashenko, P. Piskorz, I. Komaromi, R. Gomperts, R.L. Martin, D.J. Fox, T. Keith, M.A. Al-Laham, C.Y. Peng, A. Nanayakkara, C. Gonzalez, M. Challacombe, P.M.W. Gill, B. Johnson, W. Chen, M.W. Wong, J.L. Andres, C. Gonzalez, M. Head-Gordon, E.S. Replogle and J.A. Pople, *Gaussian 98, Revision A.3*. Gaussian, Pittsburgh, PA, 1998.

[106] A.P. Scott and L. Radom, *J. Phys. Chem.* 1996, **100**, 16502.

[107] J.-H. Chang, G.S. Xuan, C.J. Li and J.-H. Kim, Formation and characterization of the orientation controlled self-assembled monolayer of cytochrome *c*, *Abstracts of Papers, 223rd ACS National Meeting, Orlando, FL, April 7–11, 2002*, COLL-244.

[108] J.M. Zhang, D.H. Zhang and D.Y. Shen, Orientation study of atactic poly(methyl methacrylate) thin film by SERS and RAIR spectra, *Macromolecules* 2002, **35**, 5140–5144.

[109] T. Imae, in *Encyclopedia of Surface and Colloid Science*, Marcel Dekker, Basel, 2002, pp. 3547–3558.

[110] A. Ulman, *Ultrathin Organic Films*, Academic Press, Boston, 1991.

[111] Z. Zhang, N. Yoshida, T. Imae, Q. Xue, M. Bai, J. Jiang and Z. Liu, A self-assembled monolayer of an alkanoic acid-derivatized porphyrin on gold surface: a structural investigation by surface plasmon resonance, ultraviolet–visible, and infrared spectroscopies, *J. Colloid Interface Sci.* 2001, **243**, 382–387.

[112] Z. Zhang, T. Imae, H. Sato, A. Watanabe and Y. Ozaki, Surface-enhanced Raman scattering and surface-enhanced infrared absorption spectroscopic studies of a metalloporphyrin monolayer film formed on pyridine self-assembled monolayer-modified gold, *Langmuir* 2001, **17**, 4564–4568.

[113] C.R. Olave, E.A.F. Carrasco, M. Campos-Vallette, M.S. Saavedra, G.F. Diaz, R.E. Clavijo, W. Figueroa, J.V. Garcia-Ramos, S. Sanchez-Cortes, C. Domingo, J. Costamagna and A. Rios, Vibrational study of the interaction of dinaphthalenic Ni(II) and Cu(II) azamacrocycle complexes methyl and phenyl substituted with different metal surfaces, *Vib. Spectrosc.* 2002, **28**, 287–297.

[114] M. Campos-Vallette, M.S. Saavedra, G.F. Diaz, R.E. Clavijo, Y. Martinez, F. Mendizabal, J. Costamagna, J.C. Canales, J.V. Garcia-Ramos and S. Sanchez-Cortes, Surface-enhanced vibrational study of azabipiridyl and its Co(II), Ni(II) and Cu(II) complexes, *Vib. Spectrosc.* 2001, **27**, 15–27.

[115] H. Nagaoka and T. Imae, Poly(amido amine) dendrimer adsorption onto 3-mercaptopropionic acid self-assembled monolayer formed on Au surface–investigation by surface enhanced spectroscopy and surface plasmon sensing, *Trans. Mater. Res. Soc. Jpn.* 2001, **26**, 945–948.

[116] H. Nagaoka and T. Imae, The construction of layered architectures of dendrimers – adsorption layers of amino-terminated dendrimers on 3-mercaptopropionic acid self-assembled monolayer formed on Au, *Int. J. Nonlinear Sci. Num. Sim.* 2002, **3**, 223–227.

[117] L.-J. Wan, H. Noda, C. Wang, C.-L. Bai and M. Osawa, Controlled orientation of individual molecules by electrode potentials, *ChemPhysChem* 2001, **2**, 617–619.

[118] T. Wandlowski, K. Ataka and D. Mayer, *In situ* infrared study of 4,4′-bipyridine adsorption on thin gold films, *Langmuir* 2002, **18**, 4331–4341.

[119] H.S. Kim, S.J. Lee, N.H. Kim, J.K. Yoon, H.K. Park and K. Kim, Adsorption characteristics of 1,4-phenylene diisocyanide on gold nanoparticles: infrared and Raman spectroscopy study, *Langmuir* 2003, **19**, 6701–6710.

[120] R. Nakamura, K. Ueda and S. Sato, *In situ* observation of the photoenhanced adsorption of water on TiO_2 films by surface-enhanced IR absorption spectroscopy, *Langmuir* 2001, **17**, 2298–2300.

[121] R. Nakamura and S. Sato, Oxygen species active for photooxidation of n-decane over TiO_2 surfaces, *J. Phys. Chem. B* 2002, **106**, 5893–5896.

[122] R. Nakamura and S. Sato, Surface-enhanced IR absorption of Pt and its application to *in situ* analysis of surface species, *Langmuir* 2002, **18**, 4433–4436.

[123] S. Sato, K. Ueda, Y. Kawasaki and R. Nakamura, *In situ* IR observation of surface species during the photocatalytic decomposition of acetic acid over TiO_2 films, *J. Phys. Chem. B* 2002, **106**, 9054–9058.

[124] D. Ferri, T. Buergi and A. Baiker, Pt and Pt/Al_2O_3 thin films for investigation of catalytic solid–liquid interfaces by ATR–IR spectroscopy: CO adsorption, H_2-induced reconstruction and surface-enhanced absorption, *J. Phys. Chem. B* 2001, **105**, 3187–3195.

[125] M. Osawa, in S. Kawata (ed), *Near-Field Optics and Surface Plasmon Polaritons*, Springer-Verlag, New York, p. 163.

[126] M. Osawa, in *Workshop on Theory and Surface Measurements of Fuel Cell Catalysis*, Magleås Conference Center, Høsterkøb, Denmark, http://www.scs.uiuc.edu/wieckowski/Lyngby/Lyngby.html, 2003.

[127] H. Noda, L.-J. Wan and M. Osawa, Dynamics of adsorption and phase formation of *p*-nitrobenzoic acid at Au(111) surface in solution. A combined surface-enhanced infrared and STM study, *Phys. Chem. Chem. Phys.* 2001, **3**, 3336–3342.

[128] Y.X. Chen, A. Miki, S. Ye, H. Sakai and M. Osawa, Formate, an active intermediate for direct oxidation of methanol on Pt electrode, *J. Am. Chem. Soc.* 2003, **125**, 3680–3681.

[129] A. Miki, S. Ye and M. Osawa, 204th Meeting, The Electrochemical Society, Orlando, FL, 2003, Abstract 1431.

[130] C.E. Foster, B.P. Barham and P.J. Reid, Resonance Raman intensity analysis of chlorine dioxide dissolved in chloroform: the role of nonpolar solvation, *J. Chem. Phys.* 2001, **114**, 8492–8504.

[131] E. Sudo, Y. Esaki and M. Sugiura, Analysis of additives in a polymer by LC/IR using surface enhanced infrared absorption spectroscopy, *Bunseki Kagaku* 2001, **50**, 703–707.

[132] P.R. Griffiths, D. Heaps and A. Bjerke, presented at the Bomem–Michelson Award Symposium, Pittcon 2003, Orlando, FL, 2003.

[133] S. Sanchez-Cortes, C. Domingo, J.V. Garcia-Ramos and J.A. Aznarez, Surface-enhanced vibrational study (SEIR and SERS) of dithiocarbamate pesticides on gold films, *Langmuir* 2001, **17**, 1157–1162.

[134] A.G. Rand, J. Ye and B.C.W.S.V. Letcher, *Food Technol.* 2002, **56**, 32–38.

[135] H.Y.N. Holman, D.L. Perry and J.C. Hunter-Cevera, *J. Microbiol. Methods* 1998, **34**, 59–91.

[136] R. Aroca, R. and Bujalski, Surface enhanced vibrational spectra of thymine, *Vib. Spectrosc.* 1999, **19**, 11–21.

[137] K. Ataka and M. Osawa, *In situ* infrared study of cytosine adsorption on gold electrodes, *J. Electroanal. Chem.* 1999, **460**, 188–196.
[138] S. Pronkin and T. Wandlowski, Time-resolved *in situ* ATR–SEIRAS study of adsorption and 2D phase formation of uracil on gold electrodes, *J. Electroanal. Chem.* 2003, **550–551**, 131–147.
[139] G.I. Dovbeshko, V.I. Chegel, N.Y. Gridina, O.P. Repnytska, Y.M. Shirshov, V.P. Tryndiak, I.M. Todor and G.I. Solyanik, *Biopolymers* 2002, **67**, 470–486.
[140] G.I. Dovbeshko, V.I. Chegel, N.Y. Gridina, O.P. Repnytska, Y.M. Shirshov, V.P. Tryndiak, I.M. Todor and S.A. Zynio, Surface enhanced infrared absorption of nucleic acids on gold substrate, *Semicond. Phys. Quantum Electron. Optoelectron.* 2001, **4**, 202–206.
[141] V.F. Chekhun, G.I. Solyanik, G.I. Kulik, V.P. Tryndiak, I.N. Todor, G.I. Dovbeshko and O.P. Repnytska, The SEIRA spectroscopy data of nucleic acids and phospholipids from sensitive- and drug-resistant rat tumours, *J. Exp. Clin. Cancer Res.* 2002, **21**, 599–607.
[142] G.I. Dovbeshko, O.P. Repnytska, E.D. Obraztsova, Y.V. Shtogun and E.O. Andreev, *Semicond. Phys. Quantum Electron. Optoelectron.* 2003, **6**, 105–108.
[143] W.B. Fischer, G. Steiner and K.C.R. Salzer, *J. Mol. Struct.* 2001, **470**, 153–158.
[144] K. Ataka and J. Heberle, Electrochemically induced surface-enhanced infrared difference absorption (SEIDA) spectroscopy of a protein monolayer, *J. Am. Chem. Soc.* 2003, **125**, 4986–4987.
[145] S. Geng, J. Friedrich, J. Gahde and L. Guo, Surface-enhanced infrared absorption (SEIRA) and its use in analysis of plasma-modified surface, *J. Appl. Polym. Sci.* 1999, **71**, 1231–1237.
[146] M. Suzuki, M. Yamamoto, A. Fujishima, T. Miyazaki, H. Hisamitsu, K. Kojima and Y. Kadoma, Raman and IR studies on adsorption behavior of adhesive monomers in a metal primer for Au, Ag, Cu, and Cr surfaces, *J. Biomed. Mater. Res.* 2002, **62**, 37–45.
[147] G.V. Andrievsky, V.K. Klochkov, A.B. Bordyuh and G.I. Dovbeshko, Comparative analysis of two aqueous-colloidal solutions of C60 fullerene with help of FTIR reflectance and UV–vis spectroscopy, *Chem. Phys. Lett.* 2002, **364**, 8–17.

Index

Surface-Enhanced Vibrational Spectroscopy R. Aroca

With kind thanks to Geoffery Jones of Information Index for compilation of this index.